OSHA's Hazard Communication Standard

A Proven Written Program for Compliance

Mark McGuire Moran

Government Institutes, Inc.
Rockville, Maryland

Government Institutes, Inc.
4 Research Place, Suite 200
Rockville, Maryland 20850

Copyright © 1996 by Government Institutes. All rights reserved.

99 98 97 96   5 4 3 2 1

No part of this work may be reproduced or transmitted in any form or by any means, electronic or mechanical, including photocopying, recording, or any information storage and retrieval system, without permission in writing from the publisher. All requests for permission to reproduce material from this work should be directed to Government Institutes, Inc., 4 Research Place, Suite 200, Rockville, Maryland 20850.

The author and publisher make no representation or warranty, express or implied, as to the completeness, correctness or utility of the information in this publication. In addition, the author and publisher assume no liability of any kind whatsoever resulting from the use of or reliance upon the contents of this book.

ISBN: 0-86587-499-9

Printed in the United States of America.

TABLE OF CONTENTS

INSTRUCTIONS FOR USE OF THE HAZARD COMMUNICATION PROGRAM v

HAZARD COMMUNICATION PROGRAM 1

- I. General Company Policy 2
- II. List of Hazardous Chemicals 2
- III. Material Safety Data Sheets 3
- IV. Labels and Other Forms of Warning 3
- V. Non-Routine Tasks 4
- VI. Training 4
- VII. Contractor Employees 5
- VIII. Additional Information 5
- IX. Your Role in Job Safety 6

    Questionnaire for Use in Developing a Written Hazard Communication Program 7
        Introduction 7
            Chemical Information List 8
            Material Safety Data Sheets (MSDS) 9
            Labels 11
            Employee Information and Training 13
            Additional Provisions 14

APPENDIX A: 29 CFR 1910.1200—HAZARD COMMUNICATION 17

**APPENDIX B: THE FUNDAMENTALS OF HAZARD COMMUNICATION/
RIGHT-TO-KNOW** 45

    Introduction 47
    Coverage 48
    Origin and Purpose 49
    The Material Safety Data Sheet (MSDS) 51
    Creating the MSDS 51
    The Information to be Contained on the MSDS 53
    MSDS Accuracy and Completeness 54
    Providing Copies of the MSDS 55
    Retaining Copies of the MSDS 56
    Users of Chemicals Must Have an MSDS 57
    Evaluating the MSDS 58
    Information and Training Requirements 60
    Implementing Training Requirements 62
    Labeling Requirements 63
    The Label's Contents 64
    Placement of Labels 65
    Appropriateness of the Label's Warning 67

The Written Hazard Communication Program 68
When Outside Contractors Work on Your Premises 69
Written Hazard Evaluation Procedures 71
Trade Secret Protection 72
Preemption of Right-to-Know Laws 73
The Federal Right-to-Know Law 75
Civil Liability Aspects of Hazard Communication Laws 78
Observing the Written Precautions 79
What Employers Must Do 81

APPENDIX C: OSHA INSTRUCTION CPL 2-2.38C INSPECTION PROCEDURES FOR THE HAZARD COMMUNICATION STANDARD 29 CFR 1910.1200, 1915.99, 1917.28, 1918.90, 1926.59, AND 1928.21 83

Purpose 84
Scope 84
References 84
Cancellation 84
Action 84
Federal Program Change 85
Special Identifiers 86
Background 86
Organization of this Instruction 87
Inspection Resources 88
Inspection Guidelines 88
Classification and Grouping of Violations 110
Interface with Other Standards 111
Evaluation 113
Appendix A: Classifications and Interpretations
    of the Hazard Communication Standard (HCS) 115
Appendix B: Sample Letter, MSDS/Label Query 151
Appendix C: Hazard Evaluation Procedures 153
Appendix D: Guide for Reviewing MSDS Completeness 157

Index 161

INSTRUCTIONS FOR USE
OF THE HAZARD COMMUNICATION PROGRAM

1. The accompanying program covers everything that is required by the written program requirements of the OSHA Hazard Communication Standard. You only need to insert your company name on the cover and insert the name and office location of the person or persons in your employ who are responsible for the various aspects of the program at your workplace, plus the place or places where the following will be kept.

    a. Copies of this program
    b. List of all chemicals present in the workplace
    c. Master list of all MSDSs

    Blank spaces have been left in this program so that those things can be inserted by you. Before doing so, however, observe the instructions that follow.

2. Read every word of the Program. It explains what the standard requires. It also contains provisions that you can change or modify in order to reflect your actual hazard communication practices. A copy of the OSHA standard is attached as Appendix A. If you choose to make changes in the written program, you should check the standard to make sure you are not thereby violating an OSHA requirement.

3. If you want a more detailed explanation of hazard communication laws, read the attached 35-page explanation, entitled: *The Fundamentals of Hazard Communication/Right-to-Know*.

4. OSHA inspectors have been instructed that each company's written program should include the name of the company person (or persons) who will be responsible for ensuring implementation of the three principal provisions of the standard: MSDSs, Labelling, and Employee Training. This program assigns all three of those responsibilities to the same person, and gives him or her the title of safety and health (S&H) manager. You can assign those responsibilities to more than one person if you choose and, of course, your job titles may be different than those used in this program.

5. You must prepare (and retain) a list of all hazardous chemicals that are present on the premises. It must be readily available at all times when employees are at work so they can read it if they choose. The list must use the *same name* for each chemical that is used on the MSDS. (See the section of the program entitled "List

of Hazardous Chemicals" on page 2.) If it will be easier for you, that list can be attached to this written program—but it doesn't have to be done that way. You *cannot* avoid the separate list requirement by simply assembling all your MSDSs in one place. The provisions for MSDSs and for a list of chemicals are two separate requirements. You must have *both*.

6. Your list of all chemicals known to be present on your premises is, of course, subject to change if: you add a new product; you add its name to the list and put a copy of its MSDS in with the other MSDSs; one of your suppliers issues a *revised* MSDS, takes the old one out of circulation, and replaces it with the revised one; or you cease using a product that contains a hazardous chemical, remove its name from your list of chemicals, and take its MSDS out of circulation.

7. Once you have adapted the written Hazard Communication Program to your workplace, make a couple of extra copies. Keep one in the company's safety office. If OSHA inspects, they can review that copy. Place another copy in a notebook and keep the notebook in a place where it is accessible to employees at all times. In larger workplaces, make multiple copies and distribute them to various places where they will be readily accessible. You can adopt other alternative methods of making the Written Program accessible to employees if you choose.

8. Employees must have the opportunity to read the written program and must be told of its existence. You can post a notice or personally tell them. The notification should include words to this effect:

> *"We have adopted a hazard communication program for this plant as required by OSHA regulation. Each employee should read it and observe it."*

Then specify how they can obtain it, for example:

> *"You will find it in a notebook on the foreman's table in the lobby (between the time clock and the break room)."*

9. If it is possible to do so—and you think it appropriate—you should reproduce your written hazard communication program and distribute it to each affected employee. You will notice that it is written in a style that is addressed to them. It gives advice to them and contains work rules for them. There could be occasions when those rules could be useful to the company—provided employees are aware of them.

10. If you haven't already done so, all the MSDSs should be put in another notebook and that notebook must always be accessible to the employees. That is covered in the section of this program entitled "Material Safety Data Sheets". If different employees work with different chemicals, you can split the MSDSs up accordingly. The employees don't have to have access to *all* MSDSs—only those for the chemicals with which they work or to which they are exposed. Keeping the MSDSs in notebook form is not required by the standard. Any method of keeping them is permissible so long as they are always accessible to employees while they are at work.

11. Make sure that the list of chemicals and the MSDSs coincide. For example, if there are 48 chemicals on your list, you should have 48 MSDSs in your notebook. If a single MSDS covers more than one chemical (for example, copper and nickel), you should have two copies of that MSDS in the notebook. One for copper, one for nickel.

12. —WARNING—
    You must observe the rules and procedures included in your written Hazard Communication Program, as well as all other written documents or instruments you have adopted to implement job safety and health provisions. You must also read and observe each MSDS. Their purpose is to provide warnings and precautions. Failure to observe them can result in severe consequences. If the procedures or rules you choose to follow are *different* from those included in the Hazard Communication Program, you must make appropriate changes in that Program before you add your name to it.

ADDITIONAL INFORMATION

Moran Associates stands behind this Hazard Communication Program. If, at any time, you have a question, want additional guidance, or wish to offer a comment or suggestion, please contact *Moran Associates* by telephone at (904) 278-5155.

NOTE: Purchasers of this Hazard Communication Program will *not* be charged any fee by *Moran Associates* for answering questions relating to this Program or providing assistance in putting it into effect.

HAZARD COMMUNICATION

PROGRAM OF

_____COMPANY NAME

I. GENERAL COMPANY POLICY

The purpose of this document is:

1. To inform our employees that our company is complying with the OSHA Hazard Communication Standard by observing all its requirements, by compiling a hazardous chemicals list, by using material safety data sheets (MSDS), by ensuring that containers are labeled, and by providing appropriate training; and

2. To explain how those responsibilities are being put into effect at this workplace.

This program applies to all work operations in our company where there may be exposure to hazardous chemicals under normal working conditions or during a foreseeable emergency situation.

Our Safety and Health (S&H) manager,_____, is the program coordinator, acting as the representative of the plant manager, who has overall responsibility for the program. He will review and update the program, as necessary. Copies of the written program may be obtained from him at_____. A copy of the OSHA hazard communication standard is attached to this program as Appendix A.

Under this program, our affected employees will be informed of the contents of the Hazard Communication Standard, the hazardous properties of the chemicals with which they work, safe handling procedures, the hazards associated with non-routine tasks, such as the cleaning of reactor vessels, the hazards associated with chemicals in unlabeled pipes, and measures to take to protect them from chemicals hazards.

II. LIST OF HAZARDOUS CHEMICALS

Our safety and health manager will make a list of all hazardous chemicals used in the facility, and will update the list as necessary. Our list of chemicals identifies all of the chemicals used in our work areas. It identifies the corresponding MSDS for each chemical by using the *same* name as that used on the corresponding MSDS. For example, if the label says "DYCOTE 8" on the container, any time you want to know more about it, go to the MSDS listed under the name "DYCOTE 8". The labels of all containers shipped to us from outside will also contain the name and address of the chemical manufacturer who can be contacted if additional information is desired.

A master list of those chemicals is maintained by our safety and health manager and will either be attached to this program as Appendix B or will be available from our S&H manager at _____.

III. MATERIAL SAFETY DATA SHEETS

MSDS's provide specific information on the chemicals in use. Our safety and health manager will maintain a binder in his office with an MSDS on every hazardous chemical on our premises. Each MSDS will be a fully completed OSHA Form 174 or the equivalent.

The Safety and Health Manager will also make sure that each work site maintains an MSDS for the hazardous chemicals in that area at a place where it is readily available to employees while they are at work. If you do not know where they are located, ask your supervisor.

Our safety and health manager is responsible for acquiring and updating MSDS's. He will contact the chemical manufacturer or vendor if additional information is necessary or if an MSDS has not been supplied with an initial shipment. All new procurements for the company must be cleared by the safety and health manager.

IV. LABELS AND OTHER FORMS OF WARNING

Our safety and health manager will also ensure that all hazardous chemicals in the plant are properly labeled and updated, as necessary. The labels list at least the chemical identity, appropriate hazard warnings, and the name and address of the manufacturer, importer or other responsible party. Our S&H manager will refer to the corresponding MSDS to assist in verifying label information.

Containers of products that include any hazardous chemical that is shipped *from* our plant will be checked by the Safety and Health manager or a person assigned to that responsibility, in order to make sure all containers are properly labeled.

If there are a number of stationary containers within a work area that have similar contents and hazards, each of them need not be labeled. However, signs will be posted to convey the needed hazard information.

On our stationary process equipment, we may sometimes substitute regular process sheets, batch tickets, blend tickets, and similar written materials for container labels—but they will contain the same information as labels. If used, those written materials will be readily available to you during your work shift.

If you transfer chemicals from a labeled container to a portable container that is intended only for your immediate use, no labels are required on the portable container.

Pipes or piping systems do not have to be labeled but their contents will be described in the training sessions. If you ever have any questions about pipes or their contents, ask your supervisor.

V. NON-ROUTINE TASKS

When you are required to perform hazardous non-routine tasks (e.g., cleaning tanks, entering confined spaces, etc.), special training will be provided in order to inform you regarding the hazardous chemicals to which you might be exposed and the proper precautions to take to reduce or avoid exposure.

VI. TRAINING

Everyone who works with or is potentially exposed to hazardous chemicals will receive initial training from the S & H manager (or a person selected by that manager) on the Hazard Communication Standard and the safe use of those hazardous chemicals to which you may be exposed.

A program that uses both audiovisual materials, classroom-type training, and/or on-the-job training has been prepared for this purpose. The training program may vary among workers but every worker will be trained in the OSHA hazard communication standard and all chemicals to which he or she may be exposed while at work.

Whenever a new hazard is introduced, additional training will be provided as appropriate. Regular safety meetings may also be used to review the information presented in the initial training. Foremen and other supervisors will be extensively trained regarding hazards and appropriate protective measures so they will be available to answer questions from employees and provide daily monitoring of safe work practices. If you are ever unsure about what you should do or uncertain about the consequences of any action you plan to take, *DON'T ACT. Ask your supervisor beforehand*!

The training plan will emphasize these items:

- Summary of the standard and this written program.

- Chemical and physical properties of hazardous materials (e.g., flash point, reactivity) and methods that can be used to detect the presence or release of chemicals (including chemicals in unlabeled pipes).

- Physical hazards of chemicals (e.g., potential for fire, explosion, etc.).

- Health hazards, including signs and symptoms of exposure, associated with exposure to chemicals and any medical condition known to be aggravated by exposure to the chemical.

- Procedures to protect against hazards (e.g., personal protective equipment required, proper use, and maintenance; work practices or methods to assure proper use and handling of chemicals; and procedures for emergency response).

- Work procedures to follow to assure protection when cleaning hazardous chemical spills and leaks.

- Where MSDS's are located, how to read and interpret the information on both labels and MSDS's, and how employees may obtain additional hazard information.

The safety and health manager or designee will regularly review our employee training program and advise management on training or retraining needs. As part of the assessment of the training program, he or she may want to obtain input from employees regarding the training they have received, and their suggestions for improving it. If you have any suggestions, give them to your supervisor. He or she will see to it that they are provided to the appropriate party.

Retraining is required when the hazard changes or when a new hazard is introduced into the workplace. It is also company policy to provide training whenever it is needed to whomever needs it. If you do not think you are fully or properly trained, or if you ever feel you need additional training in any aspect of your job or your work environment, report that to your supervisor *immediately!*

VII. CONTRACTOR EMPLOYEES

Upon notification by the responsible supervisor, our safety and health manager or a person specifically designated for the purpose will provide outside contractors with notice of any chemical hazards that may be encountered in the normal course of their work on the premises, the labeling system in use, the protective measures to be taken, the safe handling procedures to be used, and the location and availability of MSDS's. Each contractor bringing chemicals on-site must provide us with the appropriate hazard information on those substances, including the labels used and the precautionary measures to be taken in working with those chemicals.

VIII. ADDITIONAL INFORMATION

All employees, or their designated representatives, can obtain further information on this written program, the hazard communication standard, applicable MSDS's, chemical information lists and any other safety or health matter that may interest or concern them at our safety and health office located at_____.

IX. YOUR ROLE IN JOB SAFETY

This company's management recognizes that safe and healthful employment requires full-time attention to numerous details, few of which can be committed to a written document as brief as this one. Our employees gain some safety instruction while they are youngsters in the home and in school. They obtain additional instruction through experience as they grow older. They get on-the-job training here and on jobs they have held before they came to work here. All of this—plus the matters covered in this program—is part of the knowledge and experience we expect our employees to utilize during their employment.

We will help you to work in a safe and healthful manner because we want you to be healthy and free from injury but we can't do it for you. Whenever you become aware of something you think is hazardous, report it to your supervisor *IMMEDIATELY!* Whenever you have any questions or doubts, *ASK QUESTIONS* of you supervisor *AT ONCE!*

You are number one and you must always look out for number one. If you do, your chances for continued good health will be greatly improved and you will be doing yourself—and many others—a big favor.

QUESTIONNAIRE FOR USE IN DEVELOPING A WRITTEN HAZARD COMMUNICATION PROGRAM

INTRODUCTION

In order to develop a written Hazard Communication Program, it must first be known how the company is handling (or intends to handle) its responsibilities under the OSHA Hazard Communication standard. Please answer the questions on the following pages and return the completed questionnaire. In addition, prepare and submit a list of all chemicals that are present in your workplace. With that information, a written Hazard Communication Program for your workplace will be prepared and returned to you.

NAME OF COMPANY:_____

ADDRESS OF WORKPLACE:_____

A copy of the written program will be available in the following locations for review by an interested employees:

_____
(Locations Where Available)

The following pages document the action we have taken regarding our chemical information list, material safety data sheets, labels, and employee information and training.

CHEMICAL INFORMATION LIST

- Our chemical information list was compiled by:

  _____
  (Title or Name and Telephone Number of Responsible Person)

- Our chemical information list is maintained by:

  _____
  (Title or Name and Telephone Number of Responsible Person)

- Employees may request access to or a copy of the list from:

  _____
  (Name)   (Title)   (Phone No.)

(1) Describe how chemicals not already on the list are added to the list within 30 days of being introduced into the workplace.

  _____
  _____
  _____

(2) Describe the procedures used to notify the employees affected by the introduction of the new substance. (Note or attach any instructions given to the purchasing department to allow control.)

  _____
  _____
  _____

(3) Describe how independent contractors are provided access to the chemical information list, prior to the commencement of their work.

  _____
  _____
  _____

Hazard Communication Program / 9

MATERIAL SAFETY DATA SHEETS (MSDS)

- **Maintenance and Updating MSDS**

    (1) The responsibility for obtaining, and maintaining the file of MSDSs has been assigned to:

    _____

    (Title or Name)   (Telephone No.)

    (2) Describe how such sheets are to be maintained (e.g., in notebooks in the work area, in computer files, on display board) and how employees can access them.

    _____
    _____
    _____

    (3) Describe the procedure that is followed when the MSDS is not received at time of receipt of the initial shipment of the material.

    _____
    _____
    _____

    (4) Describe the procedure for replacing the MSDS as new MSDSs are sent to you by your manufacturer or distributor.

    _____
    _____
    _____

    (5) If you are using any alternative to actual data sheets in the workplace (for example, a computerized database), describe the alternative method of providing the required information.

    _____
    _____
    _____

(6) Manufacturers, distributors, or employers who prepare MSDS's must describe below the procedure for updating the MSDS when new and significant health information is found.

_____
_____
_____

- **Employee Access to MSDS**

    (1) Describe how access to MSDS upon request will be provided to each employee within one working day.

    _____
    _____
    _____

    (2) Discuss how one free copy of the requested MSDS will be provided to each employee within five working days of a request.

    _____
    _____
    _____

    (3) Employees may request a copy or access to MSDS from:

    _____
    (Name)   (Location)

LABELS

- **Incoming Containers**

    (1) The responsibility for ensuring that all incoming containers are properly labeled has been assigned to:

    _____
    (Name)

    (2) All labels on incoming containers must contain:

    - The identity of the container contents
    - The manufacturer's name and address
    - A specific target organ hazard warning

    (3) The label must be legible, in English, and prominently displayed on each container.

- **In-plant Containers**

    (1) The responsibility for ensuring that all in-plant containers are properly labeled has been assigned to:

    _____
    (Name)

    (2) If an in-house system employing numbers or graphics is used, describe the system and explain how it works:

    _____

    _____

    _____

    (3) If a method other than labeling (signs, placards, process sheets, etc.) is being used to identify the contents of a *fixed* process vessel, describe your alternative method.

    _____

    _____

(4) The person responsible for labeling *in-house portable containers* is:

_____
(Name and Title)

(5) Describe your method for labeling in-house *portable* containers.

_____

_____

_____

- **Manufacturers**

    (1) Manufacturers of hazardous substances should identify the person responsible for ensuring that labels contain the proper information as required by law.

    _____
    (Name of Person Responsible)

    (2) Manufacturers, distributors, and importers should identify the person responsible for ensuring that all shipped containers are labeled.

    _____
    (Name of Person Responsible)

    (3) Manufacturers should describe procedures to review and update label information when necessary.

    _____

    _____

    _____
    (Description of Review Mechanism)

EMPLOYEE INFORMATION AND TRAINING

- The responsibility of coordinating our Hazard Communication/Right-To-Know Training has been assigned to:

  _____
  (Name)

  (1) Describe the format to be used. For example, classroom instruction, self-paced, program learning, etc. (You may want to attach a copy of your training outline to this questionnaire).

  _____
  _____
  _____

  (2) List any training materials used, et.e., a/v, handouts, etc.

  _____
  _____
  _____

  (3) Describe the elements of the training program—compare to the elements required by the OSHA standard.

  _____
  _____
  _____

  (4) Describe the procedure employed to train new employees on hazardous chemicals prior to their initial assignment.

  _____
  _____
  _____

  (5) Describe the procedure used to train employees when a new hazard is introduced into the workplace.

  _____
  _____
  _____

14 / OSHA Hazard Communications Standard

ADDITIONAL PROVISIONS

- **Hazardous Nonroutine Tasks**

    (1) Describe how employees who perform hazardous, nonroutine tasks will be given information about hazardous chemicals to which they may be exposed during non-routine activity. This information will include:

    - Protective/safety measures the employee can take

    - Measures the company has taken to lessen the hazards including ventilation, respirators, presence of another employee; and

    - Emergency procedures

    (2) Non-routine tasks performed by employees of this company are:

    Task   Hazardous Chemical

    _____
    _____
    _____
    _____
    _____
    _____

- **Chemicals in Unlabeled Pipes**

If the work activities are performed by employees in areas where chemicals are transferred through unlabeled pipes, describe how and where the employee can get information, prior to starting work regarding:

- The Chemical in the Pipes
- Potential Hazards
- Safety Precautions Which Should be Taken

(1) List any work areas with unlabeled pipes:

_____
_____
_____
_____
_____
_____

(2) In these work areas with unlabeled pipes, the employee shall contact for further information:

_____
(Person/Position)

- **Written Hazard Determination Program**

Manufacturers, distributors, importers, and employers evaluating chemicals shall also describe in writing the procedures used to determine the hazards of chemicals they evaluate in accordance with the law and regulations.

APPENDIX A: 29 CFR 1910.1200— HAZARD COMMUNICATION

§1910.1200   Hazard Communication.

(a) Purpose. (1) The purpose of this section is to ensure that the hazards of all chemicals produced or imported are evaluated, and that information concerning their hazards is transmitted to employers and employees. This transmittal of information is to be accomplished by means of comprehensive hazard communication programs, which are to include container labeling and other forms of warning, material safety data sheets and employee training.

(2) This occupational safety and health standard is intended to address comprehensively the issue of evaluating the potential hazards of chemicals, and communicating information concerning hazards and appropriate protective measures to employees, and to preempt any legal requirements of a state, or political subdivision of a state, pertaining to this subject. Evaluating the potential hazards of chemicals, and communicating information concerning hazards and appropriate protective measures to employees, may include, for example, but is not limited to, provisions for: developing and maintaining a written hazard communication program for the workplace, including lists of hazardous chemicals present; labeling of containers of chemicals in the workplace, as well as of containers of chemicals being shipped to other workplaces; preparation and distribution of material safety data sheets to employees and downstream employers; and development and implementation of employee training programs regarding hazards of chemicals and protective measures. Under section 18 of the Act, no state or political subdivision of a state may adopt or enforce, through any court or agency, any requirement relating to the issue addressed by this Federal standard, except pursuant to a Federally-approved state plan.

(b) Scope and application. (1) This section requires chemical manufacturers or importers to assess the hazards of chemicals which they produce or import, and all employers to provide information to their employees about the hazardous chemicals to which they are exposed, by means of a hazard communication program, labels and other forms of warning, material safety data sheets, and information and training. In addition, this section requires distributors to transmit the required information to employers. (Employers who do not produce or import chemicals need only focus on those parts of this rule that deal with establishing a workplace program and communicating information to their workers. Appendix E of this section is a general guide for such employers to help them determine their compliance obligations under the rule.)

(2) This section applies to any chemical which is known to be present in the workplace in such a manner that employees may be exposed under normal conditions of use or in a foreseeable emergency.

(3) This section applies to laboratories only as follows:

(i) Employers shall ensure that labels on incoming containers of hazardous chemicals are not removed or defaced;

(ii) Employers shall maintain any material safety data sheets that are received with incoming shipments of hazardous chemicals, and ensure that they are readily accessible during each workshift to laboratory employees when they are in their work areas;

(iii) Employers shall ensure that laboratory employees are provided information and training in accordance with paragraph (h) of this section, except for the location and availability of the written hazard communication program under paragraph (h)(2)(iii) of this section; and,

(iv) Laboratory employers that ship hazardous chemicals are considered to be either a chemical manufacturer or a distributor under this rule, and thus must ensure that any containers of hazardous chemicals leaving the laboratory are labeled in accordance with paragraph (f)(1) of this section, and that a material safety data sheet is provided to distributors and other employers in accordance with paragraphs (g)(6) and (g)(7) of this section.

(4) In work operations where employees only handle chemicals in sealed containers which are not opened under normal conditions of use (such as are found in marine cargo handling, warehousing, or retail sales), this section applies to these operations only as follows:

(i) Employers shall ensure that labels on incoming containers of hazardous chemicals are not removed or defaced;

(ii) Employers shall maintain copies of any material safety data sheets that are received with incoming shipments of the sealed containers of hazardous chemicals, shall obtain a material safety data sheet as soon as possible for sealed containers of hazardous chemicals received without a material safety data sheet if an employee requests the material safety data sheet, and shall ensure that the material safety data sheets are readily accessible during each work shift to employees when they are in their work area(s); and,

(iii) Employers shall ensure that employees are provided with information and training in accordance with paragraph (h) of this section (except for the location and availability of the written hazard communication program under paragraph (h)(2)(iii) of this section), to the extent necessary to protect them in the event of a spill or leak of a hazardous chemical from a sealed container.

(5) This section does not require labeling of the following chemicals:

(i) Any pesticide as such term is defined in the Federal Insecticide, Fungicide, and Rodenticide Act (7 U.S.C. 136 et seq.), when subject to the labeling requirements of that Act and labeling regulations issued under that Act by the Environmental Protection Agency;

(ii) Any chemical substance or mixture as such terms are defined in the Toxic Substances Control Act (15 U.S.C. 2601 et seq.), when subject to the labeling requirements of that Act and labeling regulations issued under that Act by the Environmental Protection Agency.

(iii) Any food, food additive, color additive, drug, cosmetic, or medical or veterinary device or product, including materials intended for use as ingredients in such products (e.g. flavors and fragrances), as such terms are defined in the Federal Food, Drug, and Cosmetic Act (21 U.S.C. 301 et seq.) or the Virus-Serum-Toxin Act of 1913 (21 U.S.C. 151 et seq.), and regulations issued under those Acts, when they are subject to the labeling requirements under those Acts by either the Food and Drug Administration or the Department of Agriculture;

(iv) Any distilled spirits (beverage alcohols), wine, or malt beverage intended for nonindustrial use, as such terms are defined in the Federal Alcohol Administration Act (27 U.S.C. 201 et seq.) and regulations issued under that Act, when subject to the labeling requirements of that Act and labeling regulations issued under that Act by the Bureau of Alcohol, Tobacco, and Firearms;

(v) Any consumer product or hazardous substance as those terms are defined in the Consumer Product Safety Act (15 U.S.C. 2051 et seq.) and Federal Hazardous Substances Act (15 U.S.C. 1261 et seq.) respectively, when subject to a consumer product safety standard or labeling requirement of those Acts, or regulations issued under those Acts by the Consumer Product Safety Commission; and,

(vi) Agricultural or vegetable seed treated with pesticides and labeled in accordance with the Federal Seed Act (7 U.S.C. 1551 et seq.) and the labeling regulations issued under that Act by the Department of Agriculture.

(6) This section does not apply to: (i) Any hazardous waste as such term is defined by the Solid Waste Disposal Act, as amended by the Resource Conservation and Recovery Act of 1976, as amended (42 U.S.C. 6901 et seq.), when subject to regulations issued under that Act by the Environmental Protection Agency;

(ii) Any hazardous substance as such term is defined by the Comprehensive Environmental Response, Compensation, and Liability Act (CERCLA)(42 U.S.C. 9601 et seq.), when subject to regulations issued under that Act by the Environmental Protection Agency;

(iii) Tobacco or tobacco products;

(iv) Wood or wood products, including lumber which will not be processed, where the chemical manufacturer or importer can establish that the only hazard they pose to employees is the potential for flammability or combustibility (wood or wood products which have been treated with a hazardous chemical covered by this standard, and wood which may be subsequently sawed or cut, generating dust, are not exempted);

(v) Articles (as that term is defined in paragraph (c) of this section);

(vi) Food or alcoholic beverages which are sold, used, or prepared in a retail establishment (such as a grocery store, restaurant, or drinking place), and foods intended for personal consumption by employees while in the workplace;

(vii) Any drug, as that term is defined in the Federal Food, Drug, and Cosmetic Act (21 U.S.C. 301 et seq.), when it is in solid, final form for direct administration to the patient (e.g., tablets or pills); drugs which are packaged by the chemical manufacturer for sale to consumers in a retail establishment (e.g., over-the-counter drugs); and drugs intended for personal consumption by employees while in the workplace (e.g., first aid supplies);

(viii) Cosmetics which are packaged for sale to consumers in a retail establishment, and cosmetics intended for personal consumption by employees while in the workplace;

(ix) Any consumer product or hazardous substance, as those terms are defined in the Consumer Product Safety Act (15 U.S.C. 2051 et seq.) and Federal Hazardous Substances Act (15 U.S.C. 1261 et seq.) respectively, where the employer can show that it is used in the workplace for the purpose intended by the chemical manufacturer or importer of the product, and the use results in a duration and frequency of exposure which is not greater than the range of exposures that could reasonably be experienced by consumers when used for the purpose intended;

(x) Nuisance particulates where the chemical manufacturer or importer can establish that they do not pose any physical or health hazard covered under this section;

(xi) Ionizing and nonionizing radiation; and,

(xii) Biological hazards.

(c) Definitions.

Article means a manufactured item other than a fluid or particle: (i) which is formed to a specific shape or design during manufacture; (ii) which has end use function(s) dependent in whole or in part upon its shape or design during end use; and (iii) which under normal conditions of use does not release more than very small quantities, e.g., minute or trace amounts of a hazardous chemical (as determined under paragraph (d) of this section), and does not pose a physical hazard or health risk to employees.

Assistant Secretary means the Assistant Secretary of Labor for Occupational Safety and Health, U.S. Department of Labor, or designee.

Chemical means any element, chemical compound or mixture of elements and/or compounds.

Chemical manufacturer means an employer with a workplace where chemical(s) are produced for use or distribution.

Chemical name means the scientific designation of a chemical in accordance with the nomenclature system developed by the International Union of Pure and Applied Chemistry (IUPAC) or the Chemical Abstracts Service (CAS) rules of nomenclature, or a name which will clearly identify the chemical for the purpose of conducting a hazard evaluation.

Combustible liquid means any liquid having a flashpoint at or above 100 °F (37.8 °C), but below 200 °F (93.3 °C), except any mixturehaving components with flashpoints of 200 °F (93.3 °C), or higher, the total volume of which make up 99 percent or more of the total volume of the mixture.

Commercial account means an arrangement whereby a retail distributor sells hazardous chemicals to an employer, generally in large quantities over time and/or at costs that are below the regular retail price.

Common name means any designation or identification such as code name, code number, trade name, brand name or generic name used to identify a chemical other than by its chemical name.

Compressed gas means:

(i) A gas or mixture of gases having, in a container, an absolute pressure exceeding 40 psi at 70 °F (21.1 °C); or

(ii) A gas or mixture of gases having, in a container, an absolute pressure exceeding 104 psi at 130 °F (54.4 °C) regardless of the pressure at 70 °F (21.1 °C); or

(iii) A liquid having a vapor pressure exceeding 40 psi at 100 °F (37.8 °C) as determined by ASTM D-323-72.

Container means any bag, barrel, bottle, box, can, cylinder, drum, reaction vessel, storage tank, or the like that contains a hazardous chemical. For purposes of this section, pipes or piping systems, and engines, fuel tanks, or other operating systems in a vehicle, are not considered to be containers.

Designated representative means any individual or organization to whom an employee gives written authorization to exercise such employee's rights under this section. A recognized or certified collective bargaining agent shall be treated automatically as a designated representative without regard to written employee authorization.

Director means the Director, National Institute for Occupational Safety and Health, U.S. Department of Health and Human Services, or designee.

Distributor means a business, other than a chemical manufacturer or importer, which supplies hazardous chemicals to other distributors or to employers.

Employee means a worker who may be exposed to hazardous chemicals under normal operating conditions or in foreseeable emergencies. Workers such as office workers or bank tellers who encounter hazardous chemicals only in non-routine, isolated instances are not covered.

Employer means a person engaged in a business where chemicals are either used, distributed, or are produced for use or distribution, including a contractor or subcontractor.

Explosive means a chemical that causes a sudden, almost instantaneous release of pressure, gas, and heat when subjected to sudden shock, pressure, or high temperature.

Exposure or exposed means that an employee is subjected in the course of employment to a chemical that is a physical or health hazard, and includes potential (e.g. accidental or possible) exposure. "Subjected" in terms of health hazards includes any route of entry (e.g. inhalation, ingestion, skin contact or absorption.)

Flammable means a chemical that falls into one of the following categories:

(i) Aerosol, flammable means an aerosol that, when tested by the method described in 16 CFR 1500.45, yields a flame projection exceeding 18 inches at full valve opening, or a flashback (a flame extending back to the valve) at any degree of valve opening;

(ii) Gas, flammable means: (A) A gas that, at ambient temperature and pressure, forms a flammable mixture with air at a concentration of thirteen (13) percent by volume or less; or

(B) A gas that, at ambient temperature and pressure, forms a range of flammable mixtures with air wider than twelve (12) percent by volume, regardless of the lower limit;

(iii) Liquid, flammable means any liquid having a flashpoint below 100°F (37.8°C), except any mixture having components with flashpoints of 100°F (37.8°C) or higher, the total of which make up 99 percent or more of the total volume of the mixture.

(iv) Solid, flammable means a solid, other than a blasting agent or explosive as defined in §1910.109(a), that is liable to cause fire through friction, absorption of moisture, spontaneous chemical change, or retained heat from manufacturing or processing, or which can be ignited readily and when ignited burns so vigorously and persistently as to create a serious hazard. A chemical shall be considered to be a flammable solid if, when tested by the method described in 16 CFR 1500.44, it ignites and burns with a self-sustained flame at a rate greater than one-tenth of an inch per second along its major axis.

Flashpoint means the minimum temperature at which a liquid gives off a vapor in sufficient concentration to ignite when tested as follows:

(i) Tagliabue Closed Tester (See American National Standard Method of Test for Flash Point by Tag Closed Tester, Z11.24-1979 (ASTM D 56-79)) for liquids with a viscosity of less than 45 Saybolt Universal Seconds (SUS) at 100°F (37.8°C), that do not contain suspended solids and do not have a tendency to form a surface film under test; or

(ii) Pensky-Martens Closed Tester (see American National Standard Method of Test for Flash Point by Pensky-Martens Closed Tester, Z11.7-1979 (ASTM D 93-79)) for liquids with a viscosity equal to or greater than 45 SUS at 100°F (37.8°C), or that contain suspended solids, or that have a tendency to form a surface film under test; or

(iii) Setaflash Closed Tester (see American National Standard Method of Test for Flash Point by Setaflash Closed Tester (ASTM D 3278-78)).

Organic peroxides, which undergo autoaccelerating thermal decomposition, are excluded from any of the flashpoint determination methods specified above.

Foreseeable emergency means any potential occurrence such as, but not limited to, equipment failure, rupture of containers, or failure of control equipment which could result in an uncontrolled release of a hazardous chemical into the workplace.

Hazardous chemical means any chemical which is a physical hazard or a health hazard.

Hazard warning means any words, pictures, symbols, or combination thereof appearing on a label or other appropriate form of warning which convey the specific physical or health hazard(s), including target organ effects, of the chemical(s) in the container(s). (See the definitions for "physical hazard" and "health hazard" to determine the hazards which must be covered.)

Health hazard means a chemical for which there is statistically significant evidence based on at least one study conducted in accordance with established scientific principles that acute or chronic health effects may occur in exposed employees. The term "health hazard" includes chemicals which are carcinogens, toxic or highly toxic agents, reproductive toxins, irritants, corrosives, sensitizers, hepatotoxins, nephrotoxins, neurotoxins, agents which act on the hematopoietic system, and agents which damage the lungs, skin, eyes, or mucous membranes. Appendix A provides further definitions and explanations of the scope of health hazards covered by this section, and Appendix B describes the criteria to be used to determine whether or not a chemical is to be considered hazardous for purposes of this standard.

Identity means any chemical or common name which is indicated on the material safety data sheet (MSDS) for the chemical. The identity used shall permit cross-references to be made among the required list of hazardous chemicals, the label and the MSDS.

Immediate use means that the hazardous chemical will be under the control of and used only by the person who transfers it from a labeled container and only within the work shift in which it is transferred.

Importer means the first business with employees within the Customs Territory of the United States which receives hazardous chemicals produced in other countries for the purpose of supplying them to distributors or employers within the United States.

Label means any written, printed, or graphic material displayed on or affixed to containers of hazardous chemicals.

Material safety data sheet (MSDS) means written or printed material concerning a hazardous chemical which is prepared in accordance with paragraph (g) of this section.

Mixture means any combination of two or more chemicals if the combination is not, in whole or in part, the result of a chemical reaction.

Organic peroxide means an organic compound that contains the bivalent -O-O-structure and which may be considered to be a structural derivative of hydrogen peroxide where one or both of the hydrogen atoms has been replaced by an organic radical.

Oxidizer means a chemical other than a blasting agent or explosive as defined in §1910.109(a), that initiates or promotes combustion in other materials, thereby causing fire either of itself or through the release of oxygen or other gases.

Physical hazard means a chemical for which there is scientifically valid evidence that it is a combustible liquid, a compressed gas, explosive, flammable, an organic peroxide, an oxidizer, pyrophoric, unstable (reactive) or water-reactive.

Produce means to manufacture, process, formulate, blend, extract, generate, emit, or repackage.

Pyrophoric means a chemical that will ignite spontaneously in air at a temperature of 130°F (54.4°C) or below.

Responsible party means someone who can provide additional information on the hazardous chemical and appropriate emergency procedures, if necessary.

Specific chemical identity means the chemical name, Chemical Abstracts Service (CAS) Registry Number, or any other information that reveals the precise chemical designation of the substance.

Trade secret means any confidential formula, pattern, process, device, information or compilation of information that is used in an employer's business, and that gives the employer an opportunity to obtain an advantage over competitors who do not know or use it. Appendix D sets out the criteria to be used in evaluating trade secrets.

Unstable (reactive) means a chemical which in the pure state, or as produced or transported, will vigorously polymerize, decompose, condense, or will become self-reactive under conditions of shocks, pressure or temperature.

Use means to package, handle, react, emit, extract, generate as a byproduct, or transfer.

Water-reactive means a chemical that reacts with water to release a gas that is either flammable or presents a health hazard.

Work area means a room or defined space in a workplace where hazardous chemicals are produced or used, and where employees are present.

Workplace means an establishment, job site, or project, at one geographical location containing one or more work areas.

(d) Hazard determination. (1) Chemical manufacturers and importers shall evaluate chemicals produced in their workplaces or imported by them to determine if they are hazardous. Employers are not required to evaluate chemicals unless they choose not to rely on the evaluation performed by the chemical manufacturer or importer for the chemical to satisfy this requirement.

(2) Chemical manufacturers, importers or employers evaluating chemicals shall identify and consider the available scientific evidence concerning such hazards. For health hazards, evidence which is statistically significant and which is based on at least one positive study conducted in accordance with established scientific principles is considered to be sufficient to establish a hazardous effect if the results of the study meet the definitions of health hazards in this section. Appendix A shall be consulted for the scope of health hazards covered, and Appendix B shall be consulted for the criteria to be followed with respect to the completeness of the evaluation, and the data to be reported.

(3) The chemical manufacturer, importer or employer evaluating chemicals shall treat the following sources as establishing that the chemicals listed in them are hazardous:

(i) 29 CFR part 1910, subpart Z, Toxic and Hazardous Substances, Occupational Safety and Health Administration (OSHA); or,

(ii) Threshold Limit Values for Chemical Substances and Physical Agents in the Work Environment, American Conference of Governmental Industrial Hygienists (ACGIH) (latest edition). The chemical manufacturer, importer, or employer is still responsible for evaluating the hazards associated with the chemicals in these source lists in accordance with the requirements of this standard.

(4) Chemical manufacturers, importers and employers evaluating chemicals shall treat the following sources as establishing that a chemical is a carcinogen or potential carcinogen for hazard communication purposes:

(i) National Toxicology Program (NTP), Annual Report on Carcinogens (latest edition);

(ii) International Agency for Research on Cancer (IARC) Monographs (latest editions); or

(iii) 29 CFR part 1910, subpart Z, Toxic and Hazardous Substances, Occupational Safety and Health Administration.

Note: The Registry of Toxic Effects of Chemical Substances published by the National Institute for Occupational Safety and Health indicates whether a chemical has been found by NTP or IARC to be a potential carcinogen.

(5) The chemical manufacturer, importer or employer shall determine the hazards of mixtures of chemicals as follows:

(i) If a mixture has been tested as a whole to determine its hazards, the results of such testing shall be used to determine whether the mixture is hazardous;

(ii) If a mixture has not been tested as a whole to determine whether the mixture is a health hazard, the mixture shall be assumed to present the same health hazards as do the components which comprise one percent (by weight or volume) or greater of the mixture, except that the mixture shall be assumed to present a carcinogenic hazard if it contains a component in concentrations of 0.1 percent or greater which is considered to be a carcinogen under paragraph (d)(4) of this section;

(iii) If a mixture has not been tested as a whole to determine whether the mixture is a physical hazard, the chemical manufacturer, importer, or employer may use whatever scientifically valid data is available to evaluate the physical hazard potential of the mixture; and,

(iv) If the chemical manufacturer, importer, or employer has evidence to indicate that a component present in the mixture in concentrations of less than one percent (or in the case of carcinogens, less than 0.1 percent) could be released in concentrations which would exceed an established OSHA permissible exposure limit or ACGIH Threshold Limit Value, or could present a health risk to employees in those concentrations, the mixture shall be assumed to present the same hazard.

(6) Chemical manufacturers, importers, or employers evaluating chemicals shall describe in writing the procedures they use to determine the hazards of the chemical they evaluate. The written procedures are to be made available, upon request, to employees, their designated representatives, the Assistant Secretary and the Director. The written description may be incorporated into the written hazard communication program required under paragraph (e) of this section.

(e) Written hazard communication program. (1) Employers shall develop, implement, and maintain at each workplace, a written hazard communication program which at least describes how the criteria specified in paragraphs (f), (g), and (h) of this section for labels and other forms of warning, material safety data sheets, and employee information and training will be met, and which also includes the following:

(i) A list of the hazardous chemicals known to be present using an identity that is referenced on the appropriate material safety data sheet (the list may be compiled for the workplace as a whole or for individual work areas); and,

(ii) The methods the employer will use to inform employees of the hazards of non-routine tasks (for example, the cleaning of reactor vessels), and the hazards associated with chemicals contained in unlabeled pipes in their work areas.

(2) Multi-employer workplaces. Employers who produce, use, or store hazardous chemicals at a workplace in such a way that the employees of other employer(s) may be exposed (for example, employees of a construction contractor working on-site) shall additionally ensure that the hazard communication programs developed and implemented under this paragraph (e) include the following:

(i) The methods the employer will use to provide the other employer(s) on-site access to material safety data sheets for each hazardous chemical the other employer(s)' employees may be exposed to while working;

(ii) The methods the employer will use to inform the other employer(s) of any precautionary measures that need to be taken to protect employees during the workplace's normal operating conditions and in foreseeable emergencies; and,

(iii) The methods the employer will use to inform the other employer(s) of the labeling system used in the workplace.

(3) The employer may rely on an existing hazard communication program to comply with these requirements, provided that it meets the criteria established in this paragraph (e).

(4) The employer shall make the written hazard communication program available, upon request, to employees, their designated representatives, the Assistant Secretary and the Director, in accordance with the requirements of 29 CFR 1910.20 (e).

(5) Where employees must travel between workplaces during a workshift, i.e., their work is carried out at more than one geographical location, the written hazard communication program may be kept at the primary workplace facility.

(f) Labels and other forms of warning. (1) The chemical manufacturer, importer, or distributor shall ensure that each container of hazardous chemicals leaving the workplace is labeled, tagged or marked with the following information:

(i) Identity of the hazardous chemical(s);

(ii) Appropriate hazard warnings; and

(iii) Name and address of the chemical manufacturer, importer, or other responsible party.

(2)(i) For solid metal (such as a steel beam or a metal casting), solid wood, or plastic items that are not exempted as articles due to their downstream use, or shipments of whole grain, the required label may be transmitted to the customer at the time of the initial shipment, and need not be included with subsequent shipments to the same employer unless the information on the label changes;

(ii) The label may be transmitted with the initial shipment itself, or with the material safety data sheet that is to be provided prior to or at the time of the first shipment; and,

(iii) This exception to requiring labels on every container of hazardous chemicals is only for the solid material itself, and does not apply to hazardous chemicals used in conjunction with, or known to be present with, the material and to which employees handling the items in transit may be exposed (for example, cutting fluids or pesticides in grains).

(3) Chemical manufacturers, importers, or distributors shall ensure that each container of hazardous chemicals leaving the workplace is labeled, tagged, or marked in accordance with this section in a manner which does not conflict with the requirements of the Hazardous Materials Transportation Act (49 U.S.C. 1801 et seq.) and regulations issued under that Act by the Department of Transportation.

(4) If the hazardous chemical is regulated by OSHA in a substance-specific health standard, the chemical manufacturer, importer, distributor or employer shall ensure that the labels or other forms of warning used are in accordance with the requirements of that standard.

(5) Except as provided in paragraphs (f)(6) and (f)(7) of this section, the employer shall ensure that each container of hazardous chemicals in the workplace is labeled, tagged or marked with the following information:

(i) Identity of the hazardous chemical(s) contained therein; and,

(ii) Appropriate hazard warnings, or alternatively, words, pictures, symbols, or combination thereof, which provide at least general information regarding the hazards of the chemicals, and which, in conjunction with the other information immediately available to employees under the hazard communication program, will provide employees with the specific information regarding the physical and health hazards of the hazardous chemical.

(6) The employer may use signs, placards, process sheets, batch tickets, operating procedures, or other such written materials in lieu of affixing labels to individual stationary process containers, as long as the alternative method identifies the containers to which it is applicable and conveys the information required by paragraph (f)(5) of this section to be on a label. The written materials shall be readily accessible to the employees in their work area throughout each work shift.

(7) The employer is not required to label portable containers into which hazardous chemicals are transferred from labeled containers, and which are intended only for the immediate use of the employee who performs the transfer. For purposes of this section, drugs which are dispensed by a pharmacy to a health care provider for direct administration to a patient are exempted from labeling.

(8) The employer shall not remove or deface existing labels on incoming containers of hazardous chemicals, unless the container is immediately marked with the required information.

(9) The employer shall ensure that labels or other forms of warning are legible, in English, and prominently displayed on the container, or readily available in the work area throughout each work shift. Employers having employees who speak other languages may add the information in their language to the material presented, as long as the information is presented in English as well.

(10) The chemical manufacturer, importer, distributor or employer need not affix new labels to comply with this section if existing labels already convey the required information.

(11) Chemical manufacturers, importers, distributors, or employers who become newly aware of any significant information regarding the hazards of a chemical shall revise the labels for the chemical within three months of becoming aware of the new information. Labels on containers of hazardous chemicals shipped after that time shall contain the new information. If the chemical is not currently produced or imported, the chemical manufacturer, importers, distributor, or employer shall add the information to the label before the chemical is shipped or introduced into the workplace again.

(g) *Material safety data sheets.* (1) Chemical manufacturers and importers shall obtain or develop a material safety data sheet for each hazardous chemical they produce or import. Employers shall have a material safety data sheet in the workplace for each hazardous chemical which they use.

(2) Each material safety data sheet shall be in English (although the employer may maintain copies in other languages as well), and shall contain at least the following information:

(i) The identity used on the label, and, except as provided for in paragraph (i) of this section on trade secrets:

(A) If the hazardous chemical is a single substance, its chemical and common name(s);

(B) If the hazardous chemical is a mixture which has been tested as a whole to determine its hazards, the chemical and common name(s) of the ingredients which contribute to these known hazards, and the common name(s) of the mixture itself; or,

(C) If the hazardous chemical is a mixture which has not been tested as a whole:

(1) The chemical and common name(s) of all ingredients which have been determined to be health hazards, and which comprise 1% or greater of the composition, except that chemicals identified as carcinogens under paragraph (d) of this section shall be listed if the concentrations are 0.1% or greater; and,

(2) The chemical and common name(s) of all ingredients which have been determined to be health hazards, and which comprise less than 1% (0.1% for carcinogens) of the mixture, if there is evidence that the ingredient(s) could be released from the mixture in concentrations which would exceed an established OSHA permissible exposure limit or ACGIH Threshold Limit Value, or could present a health risk to employees; and,

(3) The chemical and common name(s) of all ingredients which have been determined to present a physical hazard when present in the mixture;

(ii) Physical and chemical characteristics of the hazardous chemical (such as vapor pressure, flash point);

(iii) The physical hazards of the hazardous chemical, including the potential for fire, explosion, and reactivity;

(iv) The health hazards of the hazardous chemical, including signs and symptoms of exposure, and any medical conditions which are generally recognized as being aggravated by exposure to the chemical;

(v) The primary route(s) of entry;

(vi) The OSHA permissible exposure limit, ACGIH Threshold Limit Value, and any other exposure limit used or recommended by the chemical manufacturer, importer, or employer preparing the material safety data sheet, where available;

(vii) Whether the hazardous chemical is listed in the National Toxicology Program (NTP) Annual Report on Carcinogens (latest edition) or has been found to be a potential carcinogen in the International Agency for Research on Cancer (IARC) Monographs (latest editions), or by OSHA;

(viii) Any generally applicable precautions for safe handling and use which are known to the chemical manufacturer, importer or employer preparing the material safety data sheet, including appropriate hygienic practices, protective measures during repair and maintenance of contaminated equipment, and procedures for clean-up of spills and leaks;

(ix) Any generally applicable control measures which are known to the chemical manufacturer, importer or employer preparing the material safety data sheet, such as appropriate engineering controls, work practices, or personal protective equipment;

(x) Emergency and first aid procedures;

(xi) The date of preparation of the material safety data sheet or the last change to it; and,

(xii) The name, address and telephone number of the chemical manufacturer, importer, employer or other responsible party preparing or distributing the material safety data sheet, who can provide additional information on the hazardous chemical and appropriate emergency procedures, if necessary.

(3) If no relevant information is found for any given category on the material safety data sheet, the chemical manufacturer, importer or employer preparing the material safety data sheet shall mark it to indicate that no applicable information was found.

(4) Where complex mixtures have similar hazards and contents (i.e. the chemical ingredients are essentially the same, but the specific composition varies from mixture to mixture), the chemical manufacturer, importer or employer may prepare one material safety data sheet to apply to all of these similar mixtures.

(5) The chemical manufacturer, importer or employer preparing the material safety data sheet shall ensure that

(5) The chemical manufacturer, importer or employer preparing the material safety data sheet shall ensure that the information recorded accurately reflects the scientific evidence used in making the hazard determination. If the chemical manufacturer, importer or employer preparing the material safety data sheet becomes newly aware of any significant information regarding the hazards of a chemical, or ways to protect against the hazards, this new information shall be added to the material safety data sheet within three months. If the chemical is not currently being produced or imported the chemical manufacturer or importer shall add the information to the material safety data sheet before the chemical is introduced into the workplace again.

(6)(i) Chemical manufacturers or importers shall ensure that distributors and employers are provided an appropriate material safety data sheet with their initial shipment, and with the first shipment after a material safety data sheet is updated;

(ii) The chemical manufacturer or importer shall either provide material safety data sheets with the shipped containers or send them to the distributor or employer prior to or at the time of the shipment;

(iii) If the material safety data sheet is not provided with a shipment that has been labeled as a hazardous chemical, the distributor or employer shall obtain one from the chemical manufacturer or importer as soon as possible; and,

(iv) The chemical manufacturer or importer shall also provide distributors or employers with a material safety data sheet upon request.

(7)(i) Distributors shall ensure that material safety data sheets, and updated information, are provided to other distributors and employers with their initial shipment and with the first shipment after a material safety data sheet is updated;

(ii) The distributor shall either provide material safety data sheets with the shipped containers, or send them to the other distributor or employer prior to or at the time of the shipment;

(iii) Retail distributors selling hazardous chemicals to employers having a commercial account shall provide a material safety data sheet to such employers upon request, and shall post a sign or otherwise inform them that a material safety data sheet is available;

(iv) Wholesale distributors selling hazardous chemicals to employers over-the-counter may also, as an alternative to keeping a file of material safety data sheets for all hazardous chemicals they sell, provide material safety data sheets upon the request of the employer at the time of the over-the-counter purchase, and shall post a sign or otherwise inform such employers that a material safety data sheet is available;

(v) If an employer without a commercial account purchases a hazardous chemical from a retail distributor not required to have material safety data sheets on file (i.e., the retail distributor does not have commercial accounts and does not use the materials), the retail distributor shall provide the employer, upon request, with the name, address, and telephone number of the chemical manufacturer, importer, or distributor from which a material safety data sheet can be obtained;

(vi) Wholesale distributors shall also provide material safety data sheets to employers or other distributors upon request; and,

(vii) Chemical manufacturers, importers, and distributors need not provide material safety data sheets to retail distributors that have informed them that the retail distributor does not sell the product to commercial accounts or open the sealed container to use it in their own workplaces.

(8) The employer shall maintain in the workplace copies of the required material safety data sheets for each hazardous chemical, and shall ensure that they are readily accessible during each work shift to employees when they are in their work area(s). (Electronic access, microfiche, and other alternatives to maintaining paper copies of the material safety data sheets are permitted as long as no barriers to immediate employee access in each workplace are created by such options.)

(9) Where employees must travel between workplaces during a workshift, i.e., their work is carried out at more than one geographical location, the material safety data sheets may be kept at the primary workplace facility. In this situation, the employer shall ensure that employees can immediately obtain the required information in an emergency.

(10) Material safety data sheets may be kept in any form, including operating procedures, and may be designed to cover groups of hazardous chemicals in a work area where it may be more appropriate to address the hazards of a process rather than individual

hazardous chemicals. However, the employer shall ensure that in all cases the required information is provided for each hazardous chemical, and is readily accessible during each work shift to employees when they are in in their work area(s).

(11) Material safety data sheets shall also be made readily available, upon request, to designated representatives and to the Assistant Secretary, in accordance with the requirements of 29 CFR 1910.20(e). The Director shall also be given access to material safety data sheets in the same manner.

(h) Employee information and training. (1) Employers shall provide employees with effective information and training on hazardous chemicals in their work area at the time of their initial assignment, and whenever a new physical or health hazard the employees have not previously been trained about is introduced into their work area. Information and training may be designed to cover categories of hazards (e.g., flammability, carcinogenicity) or specific chemicals. Chemical-specific information must always be available through labels and material safety data sheets.

(2) Information. Employees shall be informed of:

(i) The requirements of this section;

(ii) Any operations in their work area where hazardous chemicals are present; and,

(iii) The location and availability of the written hazard communication program, including the required list(s) of hazardous chemicals, and material safety data sheets required by this section.

(3) Training. Employee training shall include at least:

(i) Methods and observations that may be used to detect the presence or release of a hazardous chemical in the work area (such as monitoring conducted by the employer, continuous monitoring devices, visual appearance or odor of hazardous chemicals when being released, etc.);

(ii) The physical and health hazards of the chemicals in the work area;

(iii) The measures employees can take to protect themselves from these hazards, including specific procedures the employer has implemented to protect employees from exposure to hazardous chemicals, such as appropriate work practices, emergency procedures, and personal protective equipment to be used; and,

(iv) The details of the hazard communication program developed by the employer, including an explanation of the labeling system and the material safety data sheet, and how employees can obtain and use the appropriate hazard information.

(i) Trade secrets. (1) The chemical manufacturer, importer, or employer may withhold the specific chemical identity, including the chemical name and other specific identification of a hazardous chemical, from the material safety data sheet, provided that:

(i) The claim that the information withheld is a trade secret can be supported;

(ii) Information contained in the material safety data sheet concerning the properties and effects of the hazardous chemical is disclosed;

(iii) The material safety data sheet indicates that the specific chemical identity is being withheld as a trade secret; and,

(iv) The specific chemical identity is made available to health professionals, employees, and designated representatives in accordance with the applicable provisions of this paragraph.

(2) Where a treating physician or nurse determines that a medical emergency exists and the specific chemical identity of a hazardous chemical is necessary for emergency or first-aid treatment, the chemical manufacturer, importer, or employer shall immediately disclose the specific chemical identity of a trade secret chemical to that treating physician or nurse, regardless of the existence of a written statement of need or a confidentiality agreement. The chemical manufacturer, importer, or employer may require a written statement of need and confidentiality agreement, in accordance with the provisions of paragraphs (i) (3) and (4) of this section, as soon as circumstances permit.

(3) In non-emergency situations, a chemical manufacturer, importer, or employer shall, upon request, disclose a specific chemical identity, otherwise permitted to be withheld under paragraph (i)(1) of this section, to a health professional (i.e. physician, industrial hygienist, toxicologist, epidemiologist, or occupational health nurse) providing medical or other

occupational health services to exposed employee(s), and to employees or designated representatives, if:

(i) The request is in writing;

(ii) The request describes with reasonable detail one or more of the following occupational health needs for the information:

(A) To assess the hazards of the chemicals to which employees will be exposed;

(B) To conduct or assess sampling of the workplace atmosphere to determine employee exposure levels;

(C) To conduct pre-assignment or periodic medical surveillance of exposed employees;

(D) To provide medical treatment to exposed employees;

(E) To select or assess appropriate personal protective equipment for exposed employees;

(F) To design or assess engineering controls or other protective measures for exposed employees; and,

(G) To conduct studies to determine the health effects of exposure.

(iii) The request explains in detail why the disclosure of the specific chemical identity is essential and that, in lieu thereof, the disclosure of the following information to the health professional, employee, or designated representative, would not satisfy the purposes described in paragraph (i)(3)(ii) of this section:

(A) The properties and effects of the chemical;

(B) Measures for controlling workers' exposure to the chemical;

(C) Methods of monitoring and analyzing worker exposure to the chemical; and,

(D) Methods of diagnosing and treating harmful exposures to the chemical;

(iv) The request includes a description of the procedures to be used to maintain the confidentiality of the disclosed information; and,

(v) The health professional, and the employer or contractor of the services of the health professional (i.e. downstream employer, labor organization, or individual employee), employee, or designated representative, agree in a written confidentiality agreement that the health professional, employee, or designated representative, will not use the trade secret information for any purpose other than the health need(s) asserted and agree not to release the information under any circumstances other than to OSHA, as provided in paragraph (i)(6) of this section, except as authorized by the terms of the agreement or by the chemical manufacturer, importer, or employer.

(4) The confidentiality agreement authorized by paragraph (I)(3)(iv) of this section:

(i) May restrict the use of the information to the health purposes indicated in the written statement of need;

(ii) May provide for appropriate legal remedies in the event of a breach of the agreement, including stipulation of a reasonable pre-estimate of likely damages; and,

(iii) May not include requirements for the posting of a penalty bond.

(5) Nothing in this standard is meant to preclude the parties from pursuing non-contractual remedies to the extent permitted by law.

(6) If the health professional, employee, or designated representative receiving the trade secret information decides that there is a need to disclose it to OSHA, the chemical manufacturer, importer, or employer who provided the information shall be informed by the health professional, employee, or designated representative prior to, or at the same time as, such disclosure.

(7) If the chemical manufacturer, importer, or employer denies a written request for disclosure of a specific chemical identity, the denial must:

(i) Be provided to the health professional, employee, or designated representative, within thirty days of the request;

(ii) Be in writing;

(iii) Include evidence to support the claim that the specific chemical identity is a trade secret;

(iv) State the specific reasons why the request is being denied; and,

(v) Explain in detail how alternative information may satisfy the specific medical or occupational health need without revealing the specific chemical identity.

(8) The health professional, employee, or designated representative whose request for information is denied under paragraph (i)(3) of this section may refer the request and the written denial of the request to OSHA for consideration.

(9) When a health professional, employee, or designated representative refers the denial to OSHA under paragraph (i)(8) of this section, OSHA shall consider the evidence to determine if:

(i) The chemical manufacturer, importer, or employer has supported the claim that the specific chemical identity is a trade secret;

(ii) The health professional, employee, or designated representative has supported the claim that there is a medical or occupational health need for the information; and,

(iii) The health professional, employee or designated representative has demonstrated adequate means to protect the confidentiality.

(10)(i) If OSHA determines that the specific chemical identity requested under paragraph (i)(3) of this section is not a bona fide trade secret, or that it is a trade secret, but the requesting health professional, employee, or designated representative has a legitimate medical or occupational health need for the information, has executed a written confidentiality agreement, and has shown adequate means to protect the confidentiality of the information, the chemical manufacturer, importer, or employer will be subject to citation by OSHA.

(ii) If a chemical manufacturer, importer, or employer demonstrates to OSHA that the execution of a confidentiality agreement would not provide sufficient protection against the potential harm from the unauthorized disclosure of a trade secret specific chemical identity, the Assistant Secretary may issue such orders or impose such additional limitations or conditions upon the disclosure of the requested chemical information as may be appropriate to assure that the occupational health services are provided without an undue risk of harm to the chemical manufacturer, importer, or employer.

(11) If a citation for a failure to release specific chemical identity information is contested by the chemical manufacturer, importer, or employer, the matter will be adjudicated before the Occupational Safety and Health Review Commission in accordance with the Act's enforcement scheme and the applicable Commission rules of procedure. In accordance with the Commission rules, when a chemical manufacturer, importer, or employer continues to withhold the information during the contest, the Administrative Law Judge may review the citation and supporting documentation in camera or issue appropriate orders to protect the confidentiality of such matters.

(12) Notwithstanding the existence of a trade secret claim, a chemical manufacturer, importer, or employer shall, upon request, disclose to the Assistant Secretary any information which this section requires the chemical manufacturer, importer, or employer to make available. Where there is a trade secret claim, such claim shall be made no later than at the time the information is provided to the Assistant Secretary so that suitable determinations of trade secret status can be made and the necessary protections can be implemented.

(13) Nothing in this paragraph shall be construed as requiring the disclosure under any circumstances of process or percentage of mixture information which is a trade secret.

(j) Effective dates. Chemical manufacturers, importers, distributors, and employers shall be in compliance with all provisions of this section by March 11, 1994.

Note: The effective date of the clarification that the exemption of wood and wood products from the Hazard Communication standard in paragraph (b)(6)(iv) only applies to wood and wood products including lumber which will not be processed, where the manufacturer or importer can establish that the only hazard they pose to employees is the potential for flammability or combustibility, and that the exemption does not apply to wood or wood products which have been treated with a hazardous chemical covered by this standard, and wood which may be subsequently sawed or cut generating dust has been stayed from March 11, 1994 to August 11, 1994.

§1910.1200, App. A

Appendix A to §1910.1200-Health Hazard Definitions (Mandatory)

Although safety hazards related to the physical characteristics of a chemical can be objectively defined in terms of testing requirements (e.g. flammability), health hazard definitions are less precise and more subjective. Health hazards may cause measurable changes in the body-such as decreased pulmonary function. These changes are generally indicated by the occurrence of signs and symptoms in the exposed employees-such as shortness of breath, a non-measurable, subjective feeling. Employees exposed to such hazards must be apprised of both the change in body function and the signs and symptoms that may occur to signal that change.

The determination of occupational health hazards is complicated by the fact that many of the effects or signs and symptoms occur commonly in non-occupationally exposed populations, so that effects of exposure are difficult to separate from normally occurring illnesses. Occasionally, a substance causes an effect that is rarely seen in the population at large, such as angiosarcomas caused by vinyl chloride exposure, thus making it easier to ascertain that the occupational exposure was the primary causative factor. More often, however, the effects are common, such as lung cancer. The situation is further complicated by the fact that most chemicals have not been adequately tested to determine their health hazard potential, and data do not exist to substantiate these effects.

There have been many attempts to categorize effects and to define them in various ways. Generally, the terms "acute" and "chronic" are used to delineate between effects on the basis of severity or duration. "Acute" effects usually occur rapidly as a result of short-term exposures, and are of short duration. "Chronic" effects generally occur as a result of long-term exposure, and are of long duration.

The acute effects referred to most frequently are those defined by the American National Standards Institute (ANSI) standard for Precautionary Labeling of Hazardous Industrial Chemicals (Z129.1-1988)- irritation, corrosivity, sensitization and lethal dose. Although these are important health effects, they do not adequately cover the considerable range of acute effects which may occur as a result of occupational exposure, such as, for example, narcosis.

Similarly, the term chronic effect is often used to cover only carcinogenicity, teratogenicity, and mutagenicity. These effects are obviously a concern in the workplace, but again, do not adequately cover the area of chronic effects, excluding, for example, blood dyscrasias (such as anemia), chronic bronchitis and liver atrophy.

The goal of defining precisely, in measurable terms, every possible health effect that may occur in the workplace as a result of chemical exposures cannot realistically be accomplished. This does not negate the need for employees to be informed of such effects and protected from them. Appendix B, which is also mandatory, outlines the principles and procedures of hazard assessment.

For purposes of this section, any chemicals which meet any of the following definitions, as determined by the criteria set forth in Appendix B are health hazards. However, this is not intended to be an exclusive categorization scheme. If there are available scientific data that involve other animal species or test methods, they must also be evaluated to determine the applicability of the HCS.7

1. Carcinogen: A chemical is considered to be a carcinogen if:

(a) It has been evaluated by the International Agency for Research on Cancer (IARC), and found to be a carcinogen or potential carcinogen; or

(b) It is listed as a carcinogen or potential carcinogen in the Annual Report on Carcinogens published by the National Toxicology Program (NTP) (latest edition); or,

(c) It is regulated by OSHA as a carcinogen.

2. Corrosive: A chemical that causes visible destruction of, or irreversible alterations in, living tissue by chemical action at the site of contact. For example, a chemical is considered to be corrosive if, when tested on the intact skin of albino rabbits by the method described by the U.S. Department of Transportation in appendix A to 49 CFR part 173, it destroys or changes irreversibly the structure of the tissue at the site of contact following an exposure period of four hours. This term shall not refer to action on inanimate surfaces.

3. Highly toxic: A chemical falling within any of the following categories:

(a) A chemical that has a median lethal dose (LD50) of 50 milligrams or less per kilogram of body weight when administered orally to albino rats weighing between 200 and 300 grams each.

(b) A chemical that has a median lethal dose (LD50) of 200 milligrams or less per kilogram of body weight when administered by continuous contact for 24 hours (or less if death occurs within 24 hours) with the bare skin of albino rabbits weighing between two and three kilograms each.

(c) A chemical that has a median lethal concentration (LC50) in air of 200 parts per million by volume or less of gas or vapor, or 2 milligrams per liter or less of mist, fume, or dust, when administered by continuous inhalation for one hour (or less if death occurs within one hour) to albino rats weighing between 200 and 300 grams each.

4. Irritant: A chemical, which is not corrosive, but which causes a reversible inflammatory effect on living tissue by chemical action at the site of contact. A chemical is a skin irritant if, when tested on the intact skin of albino rabbits by the methods of 16 CFR 1500.41 for four hours exposure or by other appropriate techniques, it results in an empirical score of five or more. A chemical is an eye irritant if so determined under the procedure listed in 16 CFR 1500.42 or other appropriate techniques.

5. Sensitizer: A chemical that causes a substantial proportion of exposed people or animals to develop an allergic reaction in normal tissue after repeated exposure to the chemical.

6. Toxic. A chemical falling within any of the following categories:

(a) A chemical that has a median lethal dose (LD50) of more than 50 milligrams per kilogram but not more than 500 milligrams per kilogram of body weight when administered orally to albino rats weighing between 200 and 300 grams each.

(b) A chemical that has a median lethal dose (LD50) of more than 200 milligrams per kilogram but not more than 1,000 milligrams per kilogram of body weight when administered by continuous contact for 24 hours (or less if death occurs within 24 hours) with the bare skin of albino rabbits weighing between two and three kilograms each.

(c) A chemical that has a median lethal concentration (LC50) in air of more than 200 parts per million but not more than 2,000 parts per million by volume of gas or vapor, or more than two milligrams per liter but not more than 20 milligrams per liter of mist, fume, or dust, when administered by continuous inhalation for one hour (or less if death occurs within one hour) to albino rats weighing between 200 and 300 grams each.

7. Target organ effects.

The following is a target organ categorization of effects which may occur, including examples of signs and symptoms and chemicals which have been found to cause such effects. These examples are presented to illustrate the range and diversity of effects and hazards found in the workplace, and the broad scope employers must consider in this area, but are not intended to be all-inclusive.

a. Hepatotoxins: Chemicals which produce liver damage
Signs & Symptoms: Jaundice; liver enlargement
Chemicals: Carbon tetrachloride; nitrosamines

b. Nephrotoxins: Chemicals which produce kidney damage
Signs & Symptoms: Edema; proteinuria
Chemicals: Halogenated hydrocarbons; uranium

c. Neurotoxins: Chemicals which produce their primary toxic effects on the nervous system
Signs & Symptoms: Narcosis; behavioral changes; decrease in motor functions
Chemicals: Mercury; carbon disulfide

d. Agents which act on the blood or hemato-poietic system: Decrease hemoglobin function; deprive the body tissues of oxygen
Signs & Symptoms: Cyanosis; loss of consciousness
Chemicals: Carbon monoxide; cyanides

e. Agents which damage the lung: Chemicals which irritate or damage pulmonary tissue
Signs & Symptoms: Cough; tightness in chest; shortness of breath
Chemicals: Silica; asbestos

f. Reproductive toxins: Chemicals which affect the reproductive capabilities including chromosomal damage (mutations) and effects on fetuses (teratogenesis)
Signs & Symptoms: Birth defects; sterility
Chemicals: Lead; DBCP

g. Cutaneous hazards: Chemicals which affect the dermal layer of the body
Signs & Symptoms: Defatting of the skin; rashes; irritation
Chemicals: Ketones; chlorinated compounds

h. Eye hazards: Chemicals which affect the eye or visual capacity
Signs & Symptoms: Conjunctivitis; corneal damage
Chemicals: Organic solvents; acids

§1910.1200, App. B

Appendix B to §1910.1200-Hazard Determination (Mandatory)

The quality of a hazard communication program is largely dependent upon the adequacy and accuracy of the hazard determination. The hazard determination requirement of this standard is performance-oriented. Chemical manufacturers, importers, and employers evaluating chemicals are not required to follow any specific methods for determining hazards, but they must be able to demonstrate that they have adequately ascertained the hazards of the chemicals produced or imported in accordance with the criteria set forth in this Appendix.

Hazard evaluation is a process which relies heavily on the professional judgment of the evaluator, particularly in the area of chronic hazards. The performance-orientation of the hazard determination does not diminish the duty of the chemical manufacturer, importer or employer to conduct a thorough evaluation, examining all relevant data and producing a scientifically defensible evaluation. For purposes of this standard, the following criteria shall be used in making hazard determinations that meet the requirements of this standard.

1. Carcinogenicity: As described in paragraph (d)(4) of this section and Appendix A of this section, a determination by the National Toxicology Program, the International Agency for Research on Cancer, or OSHA that a chemical is a carcinogen or potential carcinogen will be considered conclusive evidence for purposes of this section. In addition, however, all available scientific data on carcinogenicity must be evaluated in accordance with the provisions of this Appendix and the requirements of the rule.

2. Human data: Where available, epidemiological studies and case reports of adverse health effects shall be considered in the evaluation.

3. Animal data: Human evidence of health effects in exposed populations is generally not available for the majority of chemicals produced or used in the workplace. Therefore, the available results of toxicological testing in animal populations shall be used to predict the health effects that may be experienced by exposed workers. In particular, the definitions of certain acute hazards refer to specific animal testing results (see Appendix A).

4. Adequacy and reporting of data. The results of any studies which are designed and conducted according to established scientific principles, and which report statistically significant conclusions regarding the health effects of a chemical, shall be a sufficient basis for a hazard determination and reported on any material safety data sheet. In vitro studies alone generally do not form the basis for a definitive finding of hazard under the HCS since they have a positive or negative result rather than a statistically significant finding.

The chemical manufacturer, importer, or employer may also report the results of other scientifically valid studies which tend to refute the findings of hazard.

§1910.1200, App. C

Appendix C to §1910.1200-Information Sources (Advisory)

The following is a list of available data sources which the chemical manufacturer, importer, distributor, or employer may wish to consult to evaluate the hazards of chemicals they produce or import:

-Any information in their own company files, such as toxicity testing results or illness experience of company employees.

-Any information obtained from the supplier of the chemical, such as material safety data sheets or product safety bulletins.

-Any pertinent information obtained from the following source list (latest editions should be used):

Condensed Chemical Dictionary
Van Nostrand Reinhold Co., 135 West 50th Street, New York, NY 10020.

The Merck Index: An Encyclopedia of Chemicals and Drugs
Merck and Company, Inc., 126 E. Lincoln Ave., Rahway, NJ 07065.

IARC Monographs on the Evaluation of the Carcinogenic Risk of Chemicals to Man
Geneva: World Health Organization, International Agency for Research on Cancer, 1972-Present. (Multivolume work). Summaries are available in supplement volumes. 49 Sheridan Street, Albany, NY 12210.

Industrial Hygiene and Toxicology, by F.A. Patty
John Wiley & Sons, Inc., New York, NY (Multivolume work).

Clinical Toxicology of Commercial Products
Gleason, Gosselin, and Hodge.

Casarett and Doull's Toxicology; The Basic Science of Poisons
Doull, Klaassen, and Amdur, Macmillan Publishing Co., Inc., New York, NY.

Industrial Toxicology, by Alice Hamilton and Harriet L. Hardy
Publishing Sciences Group, Inc., Acton, MA.

Toxicology of the Eye, by W. Morton Grant
Charles C. Thomas, 301-327 East Lawrence Avenue, Springfield, IL.

Recognition of Health Hazards in Industry
William A. Burgess, John Wiley and Sons, 605 Third Avenue, New York, NY 10158.

Chemical Hazards of the Workplace
Nick H. Proctor and James P. Hughes, J.P. Lipincott Company, 6 Winchester Terrace, New York, NY 10022.

Handbook of Chemistry and Physics
Chemical Rubber Company, 18901 Cranwood Parkway, Cleveland, OH 44128.

Threshold Limit Values for Chemical Substances and Physical Agents in the Work Environment and Biological Exposure Indices with Intended Changes

American Conference of Governmental Industrial Hygienists (ACGIH), 6500 Glenway Avenue, Bldg. D-5, Cincinnati, OH 45211.

Information on the physical hazards of chemicals may be found in publications of the National Fire Protection Association, Boston, MA.

Note: The following documents may be purchased from the Superintendent of Documents, U.S. Government Printing Office, Washington, DC 20402.

Occupational Health Guidelines

NIOSH/OSHA (NIOSH Pub. No. 81-123).
NIOSH Pocket Guide to Chemical Hazards
NIOSH Pub. No. 90-117.

Registry of Toxic Effects of Chemical Substances
(Latest edition)

Miscellaneous Documents published by the National Institute for Occupational Safety and Health.

Criteria documents.

Special Hazard Reviews.

Occupational Hazard Assessments.

Current Intelligence Bulletins.

OSHA's General Industry Standards (29 CFR Part 1910)

NTP Annual Report on Carcinogens and Summary of the Annual Report on Carcinogens.
National Technical Information Service (NTIS), 5285 Port Royal Road, Springfield, VA 22161; (703) 487-4650.

Bibliographic data bases service provider
File name

Bibliographic Retrieval Services
Biosis Previews (BRS), 1200 Route 7, Latham, NY 12110.
    CA Search
    Medlars
    NTIS
    Hazardline
    American Chemical Society Journal

Excerpta Medica
IRCS Medical Science Journal
Pre-Med
Intl Pharmaceutical Abstracts
Paper Chem.

Lockheed-DIALOG Information Service, Biosis Prev. Files Inc., 3460 Hillview Avenue, Palo Alto, CA 94304.
    CA Search Files
    CAB Abstracts
    Chemical Exposure
    Chemname
    Chemsis Files
    Chemzero
    Embase Files
    Environmental Bibliographies
    Enviroline
    Federal Research in Progress
    IRL Life Science Collection
    NTIS
  Occupational Safety and Health (NIOSH)
    Paper Chem

SDC-ORBIT, SDC Information Service, CAS Files 2500 Colorado Avenue, Santa Monica, CA 90406.
    Chemdex, 2, 3
    NTIS

National Library of Medicine
Hazardous Substances Data Bank (NSDB)
Department of Health and Human Medline Files Services, Public Health Service, National Institutes of Health, Bethesda, MD 20209.
    Toxline Files
    Cancerlit
    RTECS
    Chemline

Pergamon International Information Laboratory Hazard Bulletin Corp., 1340 Old Chain Bridge Rd., McLean, VA 22101.

Questel, Inc., 1625 Eye Street, NW, CIS/ILO Suite 818, Washington, DC 20006.
    Cancernet

Chemical Information System IC Structure and NomenclatureSearch (ICIS), Bureau of National Affairs, System (SANSS)
1133 15th Street, NW, Suite 300, Washington, DC 20005.

    Acute Toxicity (RTECS)
    Clinical Toxicology of Commercial Products
    Oil and Hazardous Materials
    Technical Assistance Data System
    CCRIS
    CESARS

Occupational Health Services, 400 MSDS Plaza Drive, Secaucus, NJ 07094.
    Hazardline

---

§1910.1200, App. D

Appendix D to §1910.1200-Definition of " Trade Secret" (Mandatory)

The following is a reprint of the Restatement of Torts section 757, comment b (1939):

b. Definition of trade secret. A trade secret may consist of any formula, pattern, device or compilation of information which is used in one's business, and which gives him an opportunity to obtain an advantage over competitors who do not know or use it. It may be a formula for a chemical compound, a process of manufacturing, treating or preserving materials, a pattern for a machine or other device, or a list of customers. It differs from other secret information in a business (see s759 of the Restatement of Torts which is not included in this Appendix) in that it is not simply information as to single or ephemeral events in the conduct of the business, as, for example, the amount or other terms of a secret bid for a contract or the salary of certain employees, or the security investments made or contemplated, or the date fixed for the announcement of a new policy or for bringing out a new model or the like. A trade secret is a process or device for continuous use in the operations of the business. Generally it relates to the production of goods, as, for example, a machine or formula for the production of an article. It may, however, relate to the sale of goods or to other operations in the business, such as a code for determining discounts, rebates or other concessions in a price list or catalogue, or a list of specialized customers, or a method of bookkeeping or other office management.

Secrecy. The subject matter of a trade secret must be secret. Matters of public knowledge or of general knowledge in an industry cannot be appropriated by one as his secret. Matters which are completely disclosed by the goods which one markets cannot be his secret. Substantially, a trade secret is known only in the particular business in which it is used. It is not requisite that only the proprietor of the business know it. He may, without losing his protection, communicate it to employees involved in its use. He may likewise communicate it to others pledged to secrecy. Others may also know of it independently, as, for example, when they have discovered the process or formula by independent invention and are keeping it secret. Nevertheless, a substantial element of secrecy must exist, so that, except by the use of improper means, there would be difficulty in acquiring the information. An exact definition of a trade secret is not possible. Some factors to be considered in determining whether given information is one's trade secret are: (1) The extent to which the information is known outside of his business; (2) the extent to which it is known by employees and others involved in his business; (3) the extent of measures taken by him to guard the secrecy of the information; (4) the value of the information to him and his competitors; (5) the amount of effort or money expended by him in developing the information; (6) the ease or difficulty with which the information could be properly acquired or duplicated by others.

Novelty and prior art. A trade secret may be a device or process which is patentable; but it need not be that. It may be a device or process which is clearly anticipated in the prior art or one which is merely a mechanical improvement that a good mechanic can make. Novelty and invention are not requisite for a trade secret as they are for patentability. These requirements are essential to patentability because a patent protects against unlicensed use of the patented device or process even by one who discovers it properly through independent research. The patent monopoly is a reward to the inventor. But such is not the case with a trade secret. Its protection is not based on a policy of rewarding or otherwise encouraging the development of secret processes or devices. The protection is merely against breach of faith and reprehensible means of learning another's secret. For this limited protection it is not appropriate to require also the kind of novelty and invention which is a requisite of patentability. The nature of the secret is, however, an important factor in determining the kind of relief that is appropriate against one who is subject to liability under the rule stated in this Section. Thus, if the secret consists of a device or process which is a novel invention, one who acquires the secret wrongfully is ordinarily enjoined from further use of it and is required to account for the profits derived from his past use. If, on the other hand, the secret consists of mechanical improvements that a good mechanic can make without resort to the secret, the wrongdoer's liability may be limited to damages, and an injunction against future use of the improvements made with the aid of the secret may be inappropriate.

§1910.1200, App. E

Appendix E to §1910.1200-(Advisory)-Guidelines for Employer Compliance

The Hazard Communication Standard (HCS) is based on a simple concept-that employees have both a need and a right to know the hazards and identities of the chemicals they are exposed to when working. They also need to know what protective measures are available to prevent adverse effects from occurring. The HCS is designed to provide employees with the information they need.

Knowledge acquired under the HCS will help employers provide safer workplaces for their employees. When employers have information about the chemicals being used, they can take steps to reduce exposures, substitute less hazardous materials, and establish proper work practices. These efforts will help prevent the occurrence of work-related illnesses and injuries caused by chemicals.

The HCS addresses the issues of evaluating and communicating hazards to workers. Evaluation of chemical hazards involves a number of technical concepts, and is a process that requires the professional judgment of experienced experts. That's why the HCS is designed so that employers who simply use chemicals, rather than produce or import them, are not required to evaluate the hazards of those chemicals. Hazard determination is the responsibility of the producers and importers of the materials. Producers and importers of chemicals are then required to provide the hazard information to employers that purchase their products.

Employers that don't produce or import chemicals need only focus on those parts of the rule that deal with establishing a workplace program and communicating information to their workers. This appendix is a general guide for such employers to help them determine what's required under the rule. It does not supplant or substitute for the regulatory provisions, but rather provides a simplified outline of the steps an average employer would follow to meet those requirements.

1. Becoming Familiar With The Rule.

OSHA has provided a simple summary of the HCS in a pamphlet entitled "Chemical Hazard Communication," OSHA Publication Number 3084. Some employers prefer to begin to become familiar with the rule's requirements by reading this pamphlet. A copy may be obtained from your local OSHA Area Office, or by contacting the OSHA Publications Office at (202) 523-9667.

The standard is long, and some parts of it are technical, but the basic concepts are simple. In fact, the requirements reflect what many employers have been doing for years. You may find that you are already largely in compliance with many of the provisions, and will simply have to modify your existing programs somewhat. If you are operating in an OSHA-approved State Plan State, you must comply with the State's requirements, which may be different than those of the Federal rule. Many of the State Plan States had hazard communication or "right-to-know" laws prior to promulgation of the Federal rule. Employers in State Plan States should contact their State OSHA offices for more information regarding applicable requirements.

The HCS requires information to be prepared and transmitted regarding all hazardous chemicals. The HCS covers both physical hazards (such as flammability), and health hazards (such as irritation, lung damage, and cancer). Most chemicals used in the workplace have some hazard potential, and thus will be covered by the rule.

One difference between this rule and many others adopted by OSHA is that this one is performance-oriented. That means that you have the flexibility to adapt the rule to the needs of your workplace, rather than having to follow specific, rigid requirements. It also means that you have to exercise more judgment to implement an appropriate and effective program.

The standard's design is simple. Chemical manufacturers and importers must evaluate the hazards of the chemicals they produce or import. Using that information, they must then prepare labels for containers, and more detailed technical bulletins called material safety data sheets (MSDS).

Chemical manufacturers, importers, and distributors of hazardous chemicals are all required to provide the appropriate labels and material safety data sheets to the employers to which they ship the chemicals. The information is to be provided automatically. Every container of hazardous chemicals you receive must be labeled, tagged, or marked with the required information. Your suppliers must also send you a properly completed material safety data sheet (MSDS) at the time of the first shipment of the chemical, and with the next shipment after the MSDS is updated with new and significant information about the hazards.

You can rely on the information received from your suppliers. You have no independent duty to analyze the chemical or evaluate the hazards of it.

Employers that "use" hazardous chemicals must have a program to ensure the information is provided to exposed employees." Use" means to package, handle, react, or transfer. This is an intentionally broad scope, and includes any situation where a chemical is present in such a way that employees may be exposed under normal conditions of use or in a foreseeable emergency.

The requirements of the rule that deal specifically with the hazard communication program are found in this section in paragraphs (e), written hazard communication program; (f), labels and other forms of warning; (g), material safety data sheets; and (h), employee information and training. The requirements of these paragraphs should be the focus of your attention. Concentrate on becoming familiar with them, using paragraphs (b), scope and application, and (c), definitions, as references when needed to help explain the provisions.

There are two types of work operations where the coverage of the rule is limited. These are laboratories and operations where chemicals are only handled in sealed containers (e.g., a warehouse). The limited provisions for these workplaces can be found in paragraph (b) of this section, scope and application. Basically, employers having these types of work operations need only keep labels on containers as they are received; maintain material safety data sheets that are received, and give employees access to them; and provide information and training for employees. Employers do not have to have written hazard communication programs and lists of chemicals for these types of operations.

The limited coverage of laboratories and sealed container operations addresses the obligation of an employer to the workers in the operations involved, and does not affect the employer's duties as a distributor of chemicals. For example, a distributor may have warehouse operations where employees would be protected under the limited sealed container provisions. In this situation, requirements for obtaining and maintaining MSDSs are limited to providing access to those received with containers while the substance is in the workplace, and requesting MSDSs when employees request access for those not received with the containers. However, as a distributor of hazardous chemicals, that employer will still have responsibilities for providing MSDSs to downstream customers at the time of the first shipment and when the MSDS is updated. Therefore, although they may not be required for the employees in the work operation, the distributor may, nevertheless, have to have MSDSs to satisfy other requirements of the rule.

2. Identify Responsible Staff

Hazard communication is going to be a continuing program in your facility. Compliance with the HCS is not a "one shot deal." In order to have a successful program, it will be necessary to assign responsibility for both the initial and ongoing activities that have to be undertaken to comply with the rule. In some cases, these activities may already be part of current job assignments. For example, site supervisors are frequently responsible for on-the-job training sessions. Early identification of the responsible employees, and involvement of them in the development of your plan of action, will result in a more effective program design. Evaluation of the effectiveness of your program will also be enhanced by involvement of affected employees.

For any safety and health program, success depends on commitment at every level of the organization. This is particularly true for hazard communication, where success requires a change in behavior. This will only occur if employers understand the program, and are committed to its success, and if employees are motivated by the people presenting the information to them.

3. Identify Hazardous Chemicals in the Workplace.

The standard requires a list of hazardous chemicals in the workplace as part of the written hazard communication program. The list will eventually serve as an inventory of everything for which an MSDS must be maintained. At this point, however, preparing the list will help you complete the rest of the program since it will give you some idea of the scope of the program required for compliance in your facility.

The best way to prepare a comprehensive list is to survey the workplace. Purchasing records may also help, and certainly employers should establish procedures to ensure that in the future purchasing procedures result in MSDSs being received before a material is used in the workplace.

The broadest possible perspective should be taken when doing the survey. Sometimes people think of "chemicals" as being only liquids in containers. The HCS covers chemicals in all physical forms-liquids, solids, gases, vapors, fumes, and mists-whether they are "contained" or not. The hazardous nature of the chemical and the potential for exposure are the factors which determine whether a chemical is covered. If it's not hazardous, it's not covered. If there is no potential for exposure (e.g., the chemical is inextricably bound and cannot be released), the rule does not cover the chemical.

Look around. Identify chemicals in containers, including pipes, but also think about chemicals generated in the work operations. For example, welding fumes, dusts, and exhaust fumes are all sources of chemical exposures. Read labels provided by suppliers for hazard information. Make a list of all chemicals in the workplace that are potentially hazardous. For your own information and planning, you may also want to note on the list the location(s) of the products within the workplace, and an indication of the hazards as found on the label. This will help you as you prepare the rest of your program.

Paragraph (b) of this section, scope and application, includes exemptions for various chemicals or workplace situations. After compiling the complete list of chemicals, you should review paragraph (b) of this section to determine if any of the items can be eliminated from the list because they are exempted materials. For example, food, drugs, and cosmetics brought into the workplace for employee consumption are exempt. So rubbing alcohol in the first aid kit would not be covered.

Once you have compiled as complete a list as possible of the potentially hazardous chemicals in the workplace, the next step is to determine if you have received material safety data sheets for all of them. Check your files against the inventory you have just compiled. If any are missing, contact your supplier and request one.

It is a good idea to document these requests, either by copy of a letter or a note regarding telephone conversations. If you have MSDSs for chemicals that are not on your list, figure out why. Maybe you don't use the chemical anymore. Or maybe you missed it in your survey. Some suppliers do provide MSDSs for products that are not hazardous. These do not have to be maintained by you.

You should not allow employees to use any chemicals for which you have not received an MSDS. The MSDS provides information you need to ensure proper protective measures are implemented prior to exposure.

4. Preparing and Implementing a Hazard Communication Program. All workplaces where employees are exposed to hazardous chemicals must have a written plan which describes how the standard will be implemented in that facility. Preparation of a plan is not just a paper exercise-all of the elements must be implemented in the workplace in order to be in compliance with the rule. See paragraph (e) of this section for the specific requirements regarding written hazard communication programs. The only work operations which do not have to comply with the written plan requirements are laboratories and work operations where employees only handle chemicals in sealed containers. See paragraph (b) of this section, scope and application, for the specific requirements for these two types of workplaces.

The plan does not have to be lengthy or complicated. It is intended to be a blueprint for implementation of your program—an assurance that all aspects of the requirements have been addressed.

Many trade associations and other professional groups have provided sample programs and other assistance materials to affected employers. These have been very helpful to many employers since they tend to be tailored to the particular industry involved. You may wish to investigate whether your industry trade groups have developed such materials.

Although such general guidance may be helpful, you must remember that the written program has to reflect what you are doing in your workplace. Therefore, if you use a generic program it must be adapted to address the facility it covers. For example, the written plan must list the chemicals present at the site, indicate who is to be responsible for the various aspects of the program in your facility, and indicate where written materials will be made available to employees.

If OSHA inspects your workplace for compliance with the HCS, the OSHA compliance officer will ask to see your written plan at the outset of the inspection. In general, the following items will be considered in evaluating your program.

The written program must describe how the requirements for labels and other forms of warning, material safety data sheets, and employee information and training, are going to be met in your facility. The following discussion provides the type of information compliance officers will be looking for to decide whether these elements of the hazard communication program have been properly addressed:

A. Labels and Other Forms of Warning

In-plant containers of hazardous chemicals must be labeled, tagged, or marked with the identity of the material and appropriate hazard warnings. Chemical manufacturers, importers, and distributors are required to ensure that every container of hazardous chemicals they ship is appropriately labeled with such information and with the name and address of the producer or other responsible party. Employers purchasing chemicals can rely on the labels provided by their suppliers. If the material is subsequently transferred by the employer from a labeled container to another container, the employer will have to label that container unless it is subject to the portable container exemption. See paragraph (f) of this section for specific labeling requirements.

The primary information to be obtained from an OSHA-required label is an identity for the material, and appropriate hazard warnings. The identity is any term which appears on the label, the MSDS, and the list of chemicals, and thus links these three sources of information. The identity used by the supplier may be a common or trade name ("Black Magic Formula"), or a chemical name (1,1,1,-trichloroethane). The hazard warning is a brief statement of the hazardous effects of the chemical ("flammable," "causes lung damage"). Labels frequently contain other information, such as precautionary measures ("do not use near open flame"), but this information is provided voluntarily and is not required by the rule. Labels must be legible, and prominently displayed. There are no specific requirements for size or color, or any specified text.

With these requirements in mind, the compliance officer will be looking for the following types of information to ensure that labeling will be properly implemented in your facility:

1. Designation of person(s) responsible for ensuring labeling of in-plant containers;

2. Designation of person(s) responsible for ensuring labeling of any shipped containers;

3. Description of labeling system(s) used;

4. Description of written alternatives to labeling of in-plant containers (if used); and,

5. Procedures to review and update label information when necessary.

Employers that are purchasing and using hazardous chemicals-rather than producing or distributing them-will primarily be concerned with ensuring that every purchased container is labeled. If materials are transferred into other containers, the employer must ensure that these are labeled as well, unless they fall under the portable container exemption (paragraph (f)(7) of this section). In terms of labeling systems, you can simply choose to use the labels provided by your suppliers on the containers. These will generally be verbal text labels, and do not usually include numerical rating systems or symbols that require special training. The most important thing to remember is that this is a continuing duty-all in-plant containers of hazardous chemicals must always be labeled. Therefore, it is important to designate someone to be responsible for ensuring that the labels are maintained as required on the containers in your facility, and that newly purchased materials are checked for labels prior to use.

B. Material Safety Data Sheets

Chemical manufacturers and importers are required to obtain or develop a material safety data sheet for each hazardous chemical they produce or import. Distributors are responsible for ensuring that their customers are provided a copy of these MSDSs. Employers must have an MSDS for each hazardous chemical which they use. Employers may rely on the information received from their suppliers. The specific requirements for material safety data sheets are in paragraph (g) of this section.

There is no specified format for the MSDS under the rule, although there are specific information requirements. OSHA has developed a non-mandatory format, OSHA Form 174, which may be used by chemical manufacturers and importers to comply with the rule. The MSDS must be in English. You are entitled to receive from your supplier a data sheet which includes all of the information required under the rule. If you do not receive one automatically, you should request one. If you receive one that is obviously inadequate, with, for example, blank spaces that are not completed, you should request an appropriately completed one. If your request for a data sheet or for a corrected data sheet does not produce the information needed, you should contact your local OSHA Area Office for assistance in obtaining the MSDS.

The role of MSDSs under the rule is to provide detailed information on each hazardous chemical, including its potential hazardous effects, its physical and chemical characteristics, and recommendations for appropriate protective measures. This information should be useful to you as the employer responsible for designing protective programs, as well as to the workers. If you are not familiar with material safety data sheets and with chemical terminology, you may need to learn to use them yourself. A glossary of MSDS terms may be helpful in this regard. Generally speaking, most employers using hazardous chemicals will primarily be concerned with MSDS information regarding hazardous effects and recommended protective measures. Focus on the sections of the MSDS that are applicable to your situation.

MSDSs must be readily accessible to employees when they are in their work areas during their workshifts. This may be accomplished in many different ways. You must decide what is appropriate for your particular workplace. Some employers keep the MSDSs in a binder in a central location (e.g., in the pick-up truck on a construction site). Others, particularly in workplaces with large numbers of chemicals, computerize the information and provide access through terminals. As long as employees can get the information when they need it, any approach may be used. The employees must have access to the MSDSs themselves-simply having a system where the information can be read to them over the phone is only permitted under the mobile worksite provision, paragraph (g)(9) of this section, when employees must travel between workplaces during the shift. In this situation, they have access to the MSDSs prior to leaving the primary worksite, and when they return, so the telephone system is simply an emergency arrangement.

In order to ensure that you have a current MSDS for each chemical in the plant as required, and that employee access is provided, the compliance officers will be looking for the following types of information in your written program:

1. Designation of person(s) responsible for obtaining and maintaining the MSDSs;

2. How such sheets are to be maintained in the workplace (e.g., in notebooks in the work area(s) or in a computer with terminal access), and how employees can obtain access to them when they are in their work area during the work shift;

3. Procedures to follow when the MSDS is not received at the time of the first shipment;

4. For producers, procedures to update the MSDS when new and significant health information is found; and,

5. Description of alternatives to actual data sheets in the workplace, if used.

For employers using hazardous chemicals, the most important aspect of the written program in terms of MSDSs is to ensure that someone is responsible for obtaining and maintaining the MSDSs for every hazardous chemical in the workplace. The list of hazardous chemicals required to be maintained as part of the written program will serve as an inventory. As new chemicals are purchased, the list should be updated. Many companies have found it convenient to include on their purchase orders the name and address of the person designated in their company to receive MSDSs.

C. Employee Information and Training

Each employee who may be "exposed" to hazardous chemicals when working must be provided information and trained prior to initial assignment to work with a hazardous chemical, and whenever the hazard changes. "Exposure" or "exposed" under the rule means that "an employee is subjected to a hazardous chemical in the course of employment through any route of entry (inhalation, ingestion, skin contact or absorption, etc.) and includes potential (e.g., accidental or possible) exposure." See paragraph (h) of this section for specific requirements. Information and training may be done either by individual chemical, or by categories of hazards (such as flammability or carcinogenicity). If there are only a few chemicals in the workplace, then you may want to discuss each one individually. Where there are large numbers of chemicals, or the chemicals change frequently, you will probably want to train generally based on the hazard categories (e.g., flammable liquids, corrosive materials, carcinogens). Employees will have access to the substance-specific information on the labels and MSDSs.

Information and training is a critical part of the hazard communication program. Information regarding hazards and protective measures are provided to workers through written labels and material safety data sheets. However, through effective information and training, workers will learn to read and understand such information, determine how it can be obtained and used in their own workplaces, and understand the risks of exposure to the chemicals in their workplaces as well as the ways to protect themselves. A properly conducted training program will ensure comprehension and understanding. It is not sufficient to either just read material to the workers, or simply hand them material to read. You want to create a climate where workers feel free to ask questions. This will help you to ensure that the information is understood. You must always remember that the underlying purpose of the HCS is to reduce the incidence of chemical source illnesses and injuries. This will be accomplished by modifying behavior through the provision of hazard information and information about protective measures. If your program works, you and your workers will better understand the chemical hazards within the workplace. The procedures you establish regarding, for example, purchasing, storage, and handling of these chemicals will improve, and thereby reduce the risks posed to employees exposed to the chemical hazards involved. Furthermore, your workers' comprehension will also be increased, and proper work practices will be followed in your workplace.

If you are going to do the training yourself, you will have to understand the material and be prepared to motivate the workers to learn. This is not always an easy task, but the benefits are worth the effort. More information regarding appropriate training can be found in OSHA Publication No. 2254 which contains voluntary training guidelines prepared by OSHA's Training Institute. A copy of this document is available from OSHA's Publications Office at (202) 219-4667.

In reviewing your written program with regard to information and training, the following items need to be considered:

1. Designation of person(s) responsible for conducting training;

2. Format of the program to be used (audiovisuals, classroom instruction, etc.);

3. Elements of the training program (should be consistent with the elements in paragraph (h) of this section); and,

4. Procedure to train new employees at the time of their initial assignment to work with a hazardous chemical, and to train employees when a new hazard is introduced into the workplace.

The written program should provide enough details about the employer's plans in this area to assess whether or not a good faith effort is being made to train employees. OSHA does not expect that every worker will be able to recite all of the information about each chemical in the workplace. In general, the most important aspects of training under the HCS are to ensure that employees are aware that they are exposed to hazardous chemicals, that they know how to read and use labels and material safety data sheets, and that, as a consequence of learning this information, they are following the appropriate protective measures established by the employer. OSHA compliance officers will be talking to employees to determine if they have received training, if they know they are exposed to hazardous chemicals, and if they know where to obtain substance-specific information on labels and MSDSs.

The rule does not require employers to maintain records of employee training, but many employers choose to do so. This may help you monitor your own program to ensure that all employees are appropriately trained. If you already have a training program, you may simply have to supplement it with whatever additional information is required under the HCS. For example, construction employers that are already in compliance with the construction training standard (29 CFR 1926.21) will have little extra training to do.

An employer can provide employees information and training through whatever means are found appropriate and protective. Although there would always have to be some training on-site (such as informing employees of the location and availability of the written program and MSDSs), employee training may be satisfied in part by general training about the requirements of the HCS and about chemical hazards on the job which is provided by, for example, trade associations, unions, colleges, and professional schools. In addition, previous training, education and experience of a worker may relieve the employer of some of the burdens of informing and training that worker. Regardless of the method relied upon, however, the employer is always ultimately responsible for ensuring that employees are adequately trained. If the compliance officer finds that the training is deficient, the employer will be cited for the deficiency regardless of who actually provided the training on behalf of the employer.

D. Other Requirements

In addition to these specific items, compliance officers will also be asking the following questions in assessing the adequacy of the program:

Does a list of the hazardous chemicals exist in each work area or at a central location?

Are methods the employer will use to inform employees of the hazards of non-routine tasks outlined?

Are employees informed of the hazards associated with chemicals contained in unlabeled pipes in their work areas?

On multi-employer worksites, has the employer provided other employers with information about labeling systems and precautionary measures where the other employers have employees exposed to the initial employer's chemicals?

Is the written program made available to employees and their designated representatives?

If your program adequately addresses the means of communicating information to employees in your workplace, and provides answers to the basic questions outlined above, it will be found to be in compliance with the rule.

5. Checklist for Compliance

The following checklist will help to ensure you are in compliance with the rule:

        Obtained a copy of the rule. ____
        Read and understood the requirements. ____
        Assigned responsibility for tasks. ____
        Prepared an inventory of chemicals. ____
        Ensured containers are labeled. ____

Obtained MSDS for each chemical. ____
Prepared written program. ____
Made MSDSs available to workers. ____
Conducted training of workers. ____
Established procedures to maintain current program. ____
Established procedures to evaluate effectiveness. ____

6. Further Assistance

If you have a question regarding compliance with the HCS, you should contact your local OSHA Area Office for assistance. In addition, each OSHA Regional Office has a Hazard Communication Coordinator who can answer your questions. Free consultation services are also available to assist employers, and information regarding these services can be obtained through the Area and Regional offices as well.

The telephone number for the OSHA office closest to you should be listed in your local telephone directory. If you are not able to obtain this information, you may contact OSHA's Office of Information and Consumer Affairs at (202) 219-8151 for further assistance in identifying the appropriate contacts.

[59 FR 6170, Feb. 9, 1994, as amended at 59 FR 17479, Apr. 13, 1994]

APPENDIX B:
THE FUNDAMENTALS OF HAZARD COMMUNICATION/RIGHT-TO-KNOW

Table of Contents

Introduction 47
1. Coverage 48
2. Origin and Purpose 49
3. The Material Safety Data Sheet (MSDS) 51
4. Creating the MSDS 51
5. The Information to be Contained on the MSDS 53
6. MSDS Accuracy and Completeness 54
7. Providing Copies of the MSDS 55
8. Retaining Copies of the MSDS 56
9. Users of Chemicals Must Have an MSDS 57
10. Evaluating the MSDS 58
11. Information and Training Requirements 60
12. Implementing Training Requirements 62
13. Labeling Requirements 63
14. A Label's Contents 64
15. Placement of Labels 65
16. Appropriateness of the Label's Warning 67
17. The Written Hazard Communication 68
18. When Outside Contractors Work on Your Premises 69
19. Written Hazard Communication Program 71
20. Trade Secret Protection 72
21. Preemption of Right-to-Know Laws 73
22. The Federal Right-to-Know Law 75
23. Civil Liberty Aspects of Hazard Communication Laws 78
24. Observing the Written Precautions 79
25. What Employers Must Do 81

THE FUNDAMENTALS OF HAZARD COMMUNICATION/RIGHT-TO-KNOW

Introduction

The purpose of the OSHA Hazard Communication Standard (HCS) is to alert workers to the existence of potentially dangerous substances in the workplace and the proper means and methods to protect themselves against them. Numerous states and some local governments have enacted substantially similar laws.

HCS is completely different from ordinary OSHA standards. They obligate an employer to prevent (or minimize) workplace hazards through such means and methods as mandatory limitations upon noise levels and airborne contaminants, or guarding requirements for machinery and elevated workstations. The HCS does not impose either mandatory limitations or requirements to abate hazardous conditions. It requires rather that <u>information</u> be developed, obtained and provided. That is also true of state and local Right-to-Know laws.

Some of the state right-to-know laws are different from HCS because they have different reporting and recordkeeping requirements, education and training provisions, and deem different substances hazardous. But HCS and Right-to-Know are essentially the same in purpose and impose substantially similar requirements upon employers. Although the following discussion will focus upon HCS, much of what is said will also apply to state and local Right-to-Know laws. It should be kept in mind, however, that compliance with one will not necessarily mean compliance with the other.

1. Coverage

The OSHA hazard communication standard is probably the most significant job safety and health regulatory action ever adopted. Yet there are a surprising number of employers who have ignored it. Some of them don't know about it. Others who do know believe that it only covers hazardous chemicals and "we don't use chemicals in our business." They are wrong on both counts.

If you don't think you use hazardous chemicals in your business or don't qualify as a chemical manufacturer, you should be cautioned that it is OSHA's contention that a "hazardous chemical" includes ordinary household detergents and that a chemical manufacturer is <u>any</u> employer who supplies his customer with a product that, in normal use or a foreseeable emergency, will release or otherwise result in exposure to a hazardous chemical. If either of these apply, so does HCS.

All employers are currently obligated to observe HCS. When first adopted in 1983, it only applied to "manufacturing," which was defined by those standard industrial classification codes beginning with the digits 20 thorough 39. Expansion to all employers occurred in 1987 when OSHA adopted the standard currently in effect. <u>See</u> 52 Federal Register 31852, August 24, 1987.[1]

---

[1] This action was taken rather reluctantly. OSHA was literally pushed into HCS expansion as a result of a court ruling that OSHA did not adequately justify its 1983 decision to exclude non-manufacturers from HCS coverage. <u>United Steelworkers of America v. Auchter</u>, 763 F.2d 728 (3d Cir. 1985). The expanded standard was challenged by representatives of the construction industry (and others) but the expanded standard was upheld in November 1988. <u>Associated Builders and Contractors, Inc. v. Secretary of Labor</u>, 13 BNA OSHC 1945 (3d Cir. 1988). The contractors later sought Supreme Court review of that court ruling but that has not affected the standard's enforceability. Employers in the construction industry must observe the standard just like all other employers until there is a Supreme Court ruling to the contrary. And that kind of ruling may never come.

Consequently, any employer that uses hazardous chemicals in his business is now obligated to observe the OSHA Hazard Communication Standard. It is difficult to conceive of any employer who is not required to observe it.[2]

2. Origin and Purpose

The standard requires that: First, chemical manufacturers and importers must assess the hazards of chemicals they produce or import and furnish detailed information to their customers upon those determined to be hazardous, and second, all employers must provide that information to their employees by means of a hazard communication program, labels on containers, material safety data sheets and access to written records and documents.

The text of the standard appears in five different places in the Code of Federal Regulations (C.F.R.). Those places are listed later in footnote 3.

The standard's beginning time varied among employers but it became fully in effect for all of them on May 23, 1988.

Distributors of hazardous chemicals are also covered. They must ensure that containers they distribute are properly labeled and that a material safety data sheet (MSDS) is provided to their customers.

The impetus for the new standard was not exactly a public outcry against hazardous chemicals, but there was such an increasing number of reports and complaints from a broad spectrum of society, that state and local legislatures began their own piecemeal approach by enacting Right-to-Know laws. They, as well as a number of others, saw problems that needed to be solved

---

[2] Perhaps in order to underscore this point, OSHA's justification for the standard estimates that there are nearly 34,000 employees exposed to hazardous chemicals who work for law firms -- a place of work not ordinarily considered to be either an "industry" or a location where exposure to chemical hazards exists. 52 Fed. Reg. 31871 (1987).

which they traced to the fact that workers in a plant are often unaware of the dangerous substances with which they work, or if they are aware, they may not know what precautions they should take. They're not alone.

Frequently employers themselves are unaware of the dangerous nature of the materials in their plants. Doctors seeking to treat an employee after exposure to a chemical substance have sometimes been unable to do so because the employer is unable or unwilling to fully identify the substance. Inhabitants of communities surrounding industrial complexes do not know the nature of the chemical vapor in the air nor the possible hazards that exposure entails. Public health officials may not be able to advise them because quite often they, too, do not have the necessary information. The same goes for fire departments that have often been unable to obtain the precise information they needed to fight outbreaks of chemical fires.

While some employers have nearly always gone to great pains to educate and inform both their employees and public officials about the chemical substances in their plants, others have not. Lacking such cooperation, there was little they could do to provide protection from those harmful substances. Consequently, both worker and citizen protection against harmful substances has often been lacking.

Right-to-Know laws are designed to provide a solution to these problems. The OSHA Hazard Communication Standard, however, does not address all them. The discussion which follows in sections 3 through 17 will cover the four principal substantive provisions of the standard: 1) Preparation, distribution and use of the material safety data sheet. 2) Education and training of employees. 3) Container labels. 4) Written hazard communication program. Other aspects of the standard and some of the potential impact it could have in civil liability cases and business practices will be discussed in sections 18 through 25.

3. **The Material Safety Data Sheet (MSDS)**

An MSDS is essentially a technical bulletin, usually 2 to 4 pages in length, that contains information about a hazardous chemical or a product containing one or more hazardous chemicals, such as its composition, its chemical and physical characteristics, its health and safety hazards, and the precautions for safe handling and use.

The MSDS is well known in the maritime industry because it became a requirement under the Maritime Safety Act over 25 years ago. It has come into fairly general use throughout industry over the past decade because product manufacturers have come to regard it as good business practice to provide that kind of information on their products for their customers who request it as well as health professionals who seek such information in the course of their official responsibilities.

For example, the manufacturer of your favorite household detergent probably has a MSDS available. If it's Mr. Clean, Oakite, Lestoil, or whatever, the MSDS for that product will probably provide the detailed information regularly included on an MSDS required under the HCS. The product manufacturer is usually happy to supply the MSDS to anyone who requests it.

The MSDS is the centerpiece of the Hazard Communication Standard. Labels are keyed to it and the employee training and information requirements are based upon it. The MSDS serves as the primary vehicle for transmitting detailed hazard information to <u>both</u> employers and employees.

4. **Creating the MSDS** is the responsibility of the chemical manufacturer, importer or distributor. Other employers who use chemicals in their businesses are entitled to rely upon the information on the MSDS.

A MSDS must be prepared or obtained by the chemical manufacturer, distributor or importer for every product that contains a hazardous chemical. An exception to this requirement exists for hazardous waste, tobacco products, wood products and "articles".

Articles are defined essentially as manufactured items which are formed into a specific shape or design for a particular end-use function <u>and</u> which will <u>not</u> release or otherwise result in exposure to a hazardous chemical under normal conditions of use. For example, tools, nuts and bolts. Clothes, however, are not "articles" and it is possible that a MSDS will be required for some kinds of clothing. That may also be true for certain felt-tip pens, magic markers, and many other manufactured products. In November 1988, a U.S. Court of Appeals decision held that electrical brushes made predominantly of copper and graphite were <u>not</u> "articles" because employees who work with the brushes are exposed to copper and graphite dust on their hands and in the air. <u>General Carbon Co. v. OSAHRC</u>, 860 F.2d 479 (D.C. Cir. 1988). Thus, an MSDS had to be prepared for those brushes.

To ensure that there is a MSDS for every hazardous chemical, its manufacturer, distributor or importer must evaluate the chemicals they produce or import to determine if they are hazardous. That process is known as a "hazard determination." Subsection (d)[3]/ and Appendix B of the standard include some requirements for that and identify some substances considered hazardous <u>per se</u>,[4]/ but OSHA recognizes that the process must rely heavily upon the professional judgment of the evaluator. Further discussion of that requirement appears later. <u>See</u> section 19 of this paper.

---

[3]/ The standard is codified at five different places in the Code of Federal Regulations. The text of each of the five is identical. Therefore, the subsections and appendix references will be the same no matter which CFR text is consulted. The five places in Title 29, C.F.R., where the standard appears are: §§1910.1200, 1915.99, 1917.28, 1918.90, and 1926.59.

[4]/ Many Right-to-Know laws, in contrast, list <u>all</u> of the substances considered to be hazardous or make reference to particular lists that are so considered.

5. **The information to be contained on the MSDS** is rather extensive and it must be in English. <u>See</u> subsection (g) of the standard. It includes the following 12 categories:

(1). The chemical or common name of each included element and chemical compound that contributes to its hazardous nature. The name used must be the same as the name used on the label of the container in which the chemical is kept.[5/]

(2). The physical and chemical characteristics of the subject of the MSDS, such as vapor pressure and flash point.

(3). Its physical hazards such as its potential for fire, explosion and reactivity.

(4). Its health hazards, including signs and symptoms which exposure produces, and any medical conditions which are generally recognized to be aggravated by exposure to it.

(5). Its primary route of entry into a person's body such as breathing, ingestion or through the skin.

(6). Its OSHA permissible exposure limit (PEL) if there is one, its Threshold Limit Value (TLV), and any other exposure limit used or recommended by the preparer of the MSDS.

(7). Whether it is an actual or potential carcinogen.

(8). The generally applicable precautions for its safe

[5/] If it's a mixture that has been tested in that state, the common name of the mixture must be listed. If it has not been so tested, the chemical and common name of <u>each ingredient</u> (constituting 1% or more of the whole) that has been determined to be a <u>health</u> hazard must be listed. However, if an ingredient is identified as a carcinogen, that ingredient must be listed if the concentration is as small as 1/10 of 1%. Where the <u>mixture</u> itself poses a <u>physical</u> hazard (like fire or explosion), the ingredient responsible for that hazard must also be listed by chemical or common name, but the person making up the MSDS has much wider latitude in determining that the mixture as a whole does not pose the physical hazards of its ingredients. Consequently, there is no automatic 1% cut-off point in such cases.

handling and use, including appropriate hygiene practices, protective measures during repair and maintenance of contaminated equipment, and procedures for clean-up of spills and leaks.

(9). The generally applicable control measures such as appropriate engineering controls, work practices, or personal protective equipment.

(10). Emergency and first aid procedures.

(11). The calendar date on which the MSDS was prepared or updated.

(12). The name, address and telephone number of the manufacturer, importer or other party who prepared the MSDS. This has to be someone who can provide additional information and appropriate emergency procedures on the subject covered by the MSDS.

6. **MSDS Accuracy and Completeness**

Information on each of these 12 categories must be included on the MSDS if it is known. Where there is no such information known, the category must be listed followed by the statement that no applicable information has been found. In other words, no blanks can be left on the MSDS. The existence of numerous blanks was a common feature of many MSDS's used in the past.

There is no periodic update requirement. However, the information on the MSDS must be updated whenever its preparer becomes newly aware of any significant information regarding the hazards of the covered chemical or substance, or protective methods therefor. The updating must be done three months after that new awareness is acquired.

A single MSDS may cover several different mixtures if they pose similar hazards and ingredients (though in different amounts).

In response to a number of requests for an approved MSDS format, OSHA has produced a 2-page non-mandatory form. Some companies follow it. Many do not. Those who want to obtain it may contact any OSHA office.

Subsection (g)(5) of the standard provides that the chemical manufacturer, importer or employer who prepares the MSDS "shall ensure that the information recorded accurately reflects the scientific evidence used in making the hazard determination." Consequently, anyone preparing such a document should do so with great care. It would not be wise to copy someone else's MSDS and adopt it as your own. If the one you copied was inaccurate, *you* would be responsible.

Although OSHA makes no mention of it, that provision will almost certainly be used as grist for the mill of those who may seek damages resulting from future industrial calamities -- or those who may choose to initiate actions to eliminate potential disasters. Workers will not always be the claimants in that future litigation either. The prospect of litigation between chemical suppliers, users, manufacturers, consumers, local residents, and other interest groups cannot be lightly dismissed. Further discussion of this matter appears later in this paper.

7. **Providing copies of the MSDS** to purchasers of the hazardous chemical (or product containing a hazardous chemical) is the affirmative responsibility of the chemical manufacturer, distributor and importer. Subsection (g)(6). He cannot await a request. It must occur at the time of initial shipment although it need not be physically included in the shipment. It can be mailed separately in advance or at the same time, or it can be transmitted via computer link-ups. The same rule applies to each subsequent update of the MSDS.

If that does not happen, the purchaser is nevertheless obligated to obtain the MSDS as soon as possible.

There is no prescribed form for a MSDS. Any format is permissible so long as it is capable of serving its purpose:

permitting the required information to be readily available when needed.

8. **Retaining copies of the MSDS** is the responsibility of each employer who uses hazardous chemicals (or products containing them) in his business. Subsection (g)(8). However, there is much more to that requirement. They must be readily accessible to employees during their work shift. That means that keeping them in a file in an office that is open from 9 to 5 will not be enough if the plant is open for longer hours or the office file is not readily accessible to an employee working where the chemical is present.

Ideally, they should be allocated and located by work area. For example, if the shipping room employees are working around chemical A and the receiving room employees around chemical B, the MSDS for chemical A should be located in the shipping room and for chemical B in the receiving room. There is no requirement that _all_ employees have access to _all_ MSDS's. The work area has been defined by separate rooms in this example but that same principle applies where employees are assigned to work by product line, job function or by any other criterion. The MSDS for the chemicals to which an employee may be exposed must be readily accessible to him whenever -- and wherever -- he is at work.

This standard does not require that an employer must retain the MSDS for any product he is no longer using but, under a different OSHA regulation, 29 CFR § 1910.20(d)(1)(ii)(B), some MSDS information must be retained for at least 30 years.[6] The employer does not have to keep an MSDS in the same format in which it existed at the time of receipt. For example, it can be included as part of the operating procedures, work instructions or employee handbook.

---

[6] That is provided for in the OSHA "records access" regulation. 29 C.F.R. §1910.20.

Copies of the MSDS must also be furnished to local agencies under some circumstances. See the Federal Right-To-Know Law discussion later in this paper.

9. <u>Users of chemicals must have an MSDS</u> for each hazardous chemical or product containing a hazardous chemical. It does not matter where such a product came from or how it got on the premises. Perhaps the employer's most difficult job under the standard is not obtaining MSDS copies from the chemical manufacturers or even making them readily accessible to the employees concerned, it is finding out what hazardous chemicals are already <u>on</u> the premises -- and what is <u>brought on</u> the premises each day.

No matter how thorough one may search, you can be pretty sure that there will often be some materials or products on the employer's premises that are regulated by the standard, but for one reason or another, no MSDS has been obtained. Those oversights probably will not be discovered by an OSHA inspection either. However, it is just such materials that are frequently involved in some kind of calamity.

For example, the principal materials used in your product lines are ordered through your purchasing office. You surely will have an MSDS for each of them. Your problem arises from the line supervisor who goes to the local hardware store to obtain detergent or paint or things of that nature for a clean-up operation or some such job. It also results from outside contractors who bring chemical products on to your premises. Few, if any, responsible management representatives have any idea that this is happening. Nevertheless, the standard mandates that <u>you</u> must have a MSDS on the detergent, the paint, or whatever.

To fulfill their responsibilities under the MSDS requirements of the standard, many have found that a change in their decentralized manner of operations was necessary. A change may also be required in the way you employ outside contractors to do work on your premises, a matter that is discussed later in this paper.

There are certain to be unexpected troubles -- and maybe even some genuine horror stories -- as a result of this MSDS requirement. Take, for example, the case of the grocery store workers who got skin rashes from groceries containing preservatives or pesticides. Or the office workers overcome by fumes when an outside contractor was stripping paint from the corridor walls. Can that sort of thing happen in your business? Do you have a cafeteria in your plant? Is there hydrogen peroxide at your first aid station? How about bleach, ammonia, window cleaners? Those questions are raised simply to inspire you to broaden your perspective as you implement your OSHA HCS responsibilities.

There are, however, some consolations. For example, you don't have to have an MSDS for consumer products that are used in the same manner that an ordinary, prudent consumer would use them provided that it would not thereby result in any different kind of exposures. You also do not have to have an MSDS for whatever food, drugs or cosmetics your employees may bring on to your premises for their own use.[7]

Clothing, however, is something else. Some fabrics in common use are treated with permanent press resins which may release formaldehyde in some situations. The standard contains no exclusion or exemption for that phenomenon.

10. **Evaluating the MSDS** is not required by the standard but it could be quite important to many employers. Indeed, those who don't do this may well be inviting trouble for themselves -- big trouble.

---

[7] The exemptions from the standard are rather narrowly defined. They should be consulted before making any judgments that an MSDS is not necessary. See subsection (b)(6) of the standard.

In view of the responsibility OSHA has placed upon them for accuracy, some chemical manufacturers who generate the MSDS's have included, rather than excluded, hazard warnings in many borderline or questionable cases. Those warnings may be unjustified or overstated. That's fine for them -- they are protecting themselves -- but it may create a problem for the user of the product. Here's an example that is based upon actual experience.

A build-up of oil and grease occurs in an area beneath the factory floor which houses supporting mechanisms for some of the machinery. About once a month, employees go down there and clean it up using a detergent sold over the counter at retail establishments throughout the country. Let's call it "Cleangood." On one such occasion, two employees were overcome by fumes. OSHA investigated and later cited the company under the general duty clause. The citation was based on the fact that the MSDS on "Cleangood" showed that it contained several parts trichloroethylene and contained various hazard warnings -- such as: "Do not use in confined spaces." OSHA witnesses also appeared at the trial to further elucidate on the horrors that trichloroethylene can produce. Surely you can get the picture. An accident occurred. A perfectly safe household detergent was thought to be the cause. Its MSDS was produced and it contained various hazard warnings. Therefore, it shouldn't have been used in the manner that the company used it. At least, that was the basis of the claim against the company.[8]

There have been similar scenarios where the alleged culprit was paint, the kind you may have used to paint the bedroom or the kitchen. Its MSDS would scare the bravest among us from coming within three counties of an open paint can. It is understandable

---

[8] an even more troublesome problem for the employer could occur if a civil suit had been brought by the injured employees. That prospect is discussed later.

why the chemical manufacturer does these things -- but the user should not be forced to bear the consequences.

There may be a solution to this problem. It's worth looking for, anyhow. Here's what to do. Carefully read the MSDS on every chemical and product that comes to your premises. If they contain hazard warnings that are not being observed, cannot be observed, or make no sense at all -- consider an alternate source of supply. The competitor's MSDS may not be such a problem for you.

Bear in mind that, although relatively few people in the past even knew of the existence of the MSDS, and even fewer paid strict attention to what it said, that's all changing now. A lot of people in the future will be scrutinizing the MSDS and comparing it against your actual practices. Then, there are sure to be second guessers. They won't read the MSDS until trouble has struck. Then they will point out that if you had paid attention to the MSDS, there would have been no trouble. Then comes the multi-million dollar lawsuit -- with you as defendant.

You should try to avoid prospective problems of this nature. They can mushroom into major episodes. Be prepared. Do your homework. Now!

11. Information and Training Requirements

The warnings and control procedures contained on the MSDS cannot be in sufficient specifics to cover the actual use situation in each workplace. That is why the standard includes employee education and training requirements. The employer must provide training that explains and reinforces the information presented to employees through the material safety data sheets and container labels. Those written materials can be made more meaningful - and the objectives of the standard more attainable - when employees understand the printed information and are aware of the actions to be taken to avoid or minimize exposure and the occurrence of adverse effects.

The standard sets forth only the minimum requirements for the required training.[9/] It is the employer's responsibility to match the nature and content of the training to the job, the employee, and the particular chemicals involved.

The four basic minimums upon which training must be given are:

1. The methods and observations that may be used to detect the presence or release of a hazardous chemical in the work area, such as monitoring devices, odor, visual appearance, etc.

2. Both the physical and health hazards of the particular chemicals to which such employee may be exposed.

3. The measures the employee can take to protect himself from these hazards, like evacuation or other emergency procedures, protective clothing and equipment, etc.

4. The details of the written hazard communication program which each employer is required to adopt pursuant to subsection (e) of the standard (discussed later in this paper). That includes an explanation of container labels, MSDS's and how employees can obtain and use the appropriate hazard information.

In addition, employees must be furnished _information_ about the operations in their own work area where hazardous chemicals are present, the location and availability of your written hazard communication program (discussed later), and the 4 minimal OSHA training requirements discussed above. For the most part, those employee information provisions are duplicative of the training program. They need not be treated separately.

[9/] See subsection (h) of the standard.

Training can be a one time only event so long as no new hazardous chemicals are introduced to the work area.[10] If that happens, appropriate training on the new chemical must be given.

Although this discussion has been phrased in terms of individual hazardous chemicals, of which there are many thousands, the employee who simultaneously works with a number of them need not be trained for each of them individually. The training can be keyed to the hazards of the process or operation rather than to the specific chemicals involved. If there is a need for information on each such chemical, the MSDS can be consulted.

12. Implementing Training Requirements

Some union leaders maintain that the training should include some kind of scare tactics that will force the employee to sit up and take notice of the potential dangers involved. That approach could also produce some unexpected and adverse consequences. Some might faint. Others might quit.

Employee motivation, education and training is a highly developed discipline. There are many expert practitioners in the field, some of whom may already be employed in your business. They certainly should be consulted concerning the nature and conduct of the training program required under the standard.

In recent years a veritable cottage industry has come into existence to furnish employers with pre-packaged training programs. Some are much too comprehensive for your particular business. Others are not comprehensive enough. It may not be wise to totally rely on them.

The best kind of training program is one that is individually tailored to each particular aspect of your business

---

[10] It should be noted, however, that _annual_ or _periodic_ training requirements are included in some state and local Right-to-Know laws.

and carried into effect at the lowest possible level of your operations. If you must have this done through outside consultants, find someone who will make house calls.

It is also important that you document your training programs even though it is not required by the standard.[11] You should make a record of the date, time and subject of each training session and maintain an attendance roster. Because many people have short memories and OSHA checks for compliance with the training requirements by asking your employees about it, many enlightened employers are obtaining signed acknowledgements from each employee who receives training. Those can prove to be quite valuable for future reference. Some employers videotape their training programs and the employees in attendance.

13. Labeling Requirements

The HCS standard had its genesis as a rather pervasive labeling regulation under which extensive information on just about everything remotely hazardous would have to appear on all containers including pipes and piping systems. It is noteworthy, therefore, at least in the opinion of some people, that it now exists with rather minimal and innocuous labeling requirements. Pipes, for one thing, do not have to be labeled. That change alone chopped off nearly two-thirds of the $2.6 billion initial cost estimate made by the Carter Administration's OSHA leadership when they originally published the proposal on January 16, 1981 -- four days before they left government service.

The substantive aspects of the standard, like a three-legged stool, have three principal supports (the MSDS, employee

---

[11] That principle should also be followed in other areas of OSHA compliance. During an inspection, OSHA will ask employees about their training. They will often forget. Some may even misrepresent the truth in an effort to put their employer in a bad light. The only way to protect against problems of that nature is to document who was trained, the subjects covered in the training, and the calendar dates thereof.

training, and container labels). Each of these is necessary to each other and essential to the accomplishment of the standard's purposes. If any one of the three is overlooked, those objectives will suffer, though perhaps they wouldn't collapse, as would the hypothetical stool.

Each container of hazardous chemicals that leaves the place of business of a chemical manufacturer, distributor or importer must contain a legible, prominently-displayed, English-language label.

Three things must appear on that label: 1. The identity of the hazardous chemical contained therein, which does not necessarily mean its official chemical name. It could be its popular name -- i.e., "Cleangood" or "Formula 64-40." 2. An appropriate hazard warning. 3. The name and address of the chemical manufacturer, importer or other responsible party.

14. The Label's Contents

The purpose of the first of these three requirements is to serve as a link to the more detailed information contained upon the MSDS. The label is not intended to be the sole, or the most complete source of information on the hazardous nature of the container's contents. In addition to these three positive requirements, there are two prohibitions: The label cannot conflict with either the requirements of the Hazardous Materials Transportation Act and its implementing regulations, or the labeling provisions of any substance-specific OSHA standard that covers the same chemical.[12]

For all other employers who have containers of substances covered by the standard (i.e., those that are *in* the workplace as contrasted with those leaving the workplace and entering the

---

[12] One such substance-specific requirement appears in 29 C.F.R. § 1910.1029(1), which specifies label requirements for coke oven emissions.

stream of commerce), the label need list only the first two of the three matters stated above.

In the preamble to the standard, OSHA repeatedly emphasized that the purpose of a label is simply to serve as an immediate warning and as a reminder of the more detailed information listed upon the MSDS and provided as part of the standard's employee training requirements. Indeed, when it originally promulgated the standard, OSHA repeatedly cautioned against "information overload" when labeling. OSHA's message seemed to be clear: Keep the labels brief and to the point.

OSHA later changed its tune on "information overload." In the General Carbon case mentioned above, OSHA argued that all hazards listed on the MSDS must also be included on the label. The court agreed with that position and held that it was a "reasonable interpretation" to hold that:

> "[A] chemical which may be dangerous in any concentration and under any circumstances is a hazardous chemical under all circumstances - that a manufacturer's duty to label is to be based upon a worst-case scenario."[13]

Id., 860 F.2d at 487, emphasis by the court.

15. Placement of Labels

Some employers say that the problem with the labeling requirement is not what has to be on the label -- but what the label has to be on. The short answer is that it has to be on each "container," a term defined in the standard as follows:

> "[A]ny bag, barrel, bottle, box, can, cylinder, drum, reaction vessel, storage

---

[13] Paradoxically, that position was taken by OSHA only a few months after it had stated in the Federal Register that: "the more detail there is on a label, the less likely it is that employees will read and act on the information." 52 Fed. Reg. 31864, August 24, 1987.

tank or the like that contains a hazardous chemical."[14]

Pipes and piping systems are *not* considered to be containers for purposes of this requirement.

Three other exceptions to the container-labeling provision are as follows:

1. Portable containers into which hazardous chemicals are transferred from labeled containers don't have to be labeled if that is done only for the immediate use of the employee who performs the transfer. That means that it will be under his control, and used only by him, during the same work shift in which it is placed in the portable container.

2. An alternative method of labeling is authorized for stationary and permanent or fixed containers. Signs, placards, process sheets, batch tickets, operating procedures, or other such written material may be used in lieu of a label so long as they identify the applicable container and include the two things necessary to all labels: Identity of the hazardous chemicals and appropriate hazard warnings.

3. Containers used in laboratories that are part of manufacturing facilities do not have to be labeled. However, when a container comes into the laboratory already bearing a label, it must be left there. It cannot be removed or defaced.

There are also four exclusions from the labeling requirement because containers bearing these substances are not subject to OSHA regulation:

1. Pesticides labeled pursuant to EPA regulation.

---

[14] Where a solid metal product such as a steel beam or a casting is not exempted as an article, the required label may be transmitted to the customer at the time of initial shipment and need not be included with subsequent shipments to the same customer unless the labeling information changes. Subsection (f)(2) of the standard.

2. Substances labeled under Food and Drug Administration requirements. Those include food, food additives, color additives, drugs and cosmetics, as well as materials intended for use in such products (e.g., flavors and fragrances).

3. Beer, ale, wine, and beverage alcohols intended for nonindustrial use and subject to labeling rules of the Bureau of Alcohol, Tobacco and Firearms.

4. Consumer products and hazardous substances that are subject to consumer product safety standards or labeling requirements of the Consumer Product Safety Commission.

Note, however, that these exclusions are from the labeling requirement only. They do not apply to the MSDS and training provisions discussed above.

It should be kept in mind that containers with labels indicating the presence of hazardous substance therein can present some real problems -- even if they are empty or have been re-filled with some perfectly harmless material.

You would be wise to make a careful survey of your workplace. Get rid of labeled containers that don't contain the substance listed on the label but, if you dispose of containers bearing your company name, remember that someone else might use them to store hazardous waste. You could be wrongfully suspected as the culprit when the TV cameras focus on the label when reporting the latest community disaster story -- or when EPA attempts to fix liability for hazardous waste. You may want to take action to protect yourself from that kind of problem.

16. Appropriateness of the Label's Warning

As stated above, subsection (f) of the standard provides that the label must contain: "Appropriate hazard warnings." There is no precise meaning of "appropriate." Presumably that will be the subject of much litigation in the future. When OSHA was asked for more specific details on that requirement, it stated that:

> "The terms 'physical' and 'health' hazards are already defined in the rule, and these are the specific hazards that are to be 'conveyed' in an 'appropriate' hazard warning. There are some situations where the specific target organ effect is not known. Where this is the case, a more general warning statement would be permitted. For example, if the only information available is an LC_{50} test result, 'harmful if inhaled' may be the only type of statement supported by the data and thus may be appropriate."

52 Fed. Reg. 31864 (1987).

Clearly, however, OSHA wants the label warning to say more than "avoid inhalation." It believes that the <u>effects</u> of inhalation should be stated, such as, "inhalation can result in chronic bronchitis or other chronic respiratory diseases."

The OSHA guidance presently available on this issue is very thin. It suffers from the same deficiencies as the recordkeeping guidance -- another ambiguous OSHA regulation with which employers must contend.

17. The Written Hazard Communication Program

Each employer must develop and implement a "hazard communication program." It must be in writing and it must be readily accessible to employees -- and the OSHA inspector.

Three things must be included in your program:

1. A list of the hazardous chemicals known to be present. It must be cross-referenced to the appropriate MSDS. It may be compiled for the workplace as a whole or broken up by work areas, production lines, etc.

2. The methods to be used to inform your employees of non-routine tasks like cleaning of reactor vessels and the means you will use to apprise employees of the hazards of the chemicals in pipes and piping systems, including protective measures they can take in the event of exposure situations.

3. The methods to be used to advise outside contractors whose employees are working on

your premises of the hazardous chemicals to which they may be exposed while they are there, and any suggestions for appropriate protective measures.

Some people maintain that all three of these matters could be covered by a two sentence Notice to Employees (which you will also give to your contractor's employees when they first arrive on your premises): 1. Read all the MSDS's in your work area. 2. Ask questions if you don't understand. OSHA inspectors have different ideas on this, however.

An actual Hazard Communication Program that seems to meet all the requirements included in subsection (e) of the standard is included as an attachment to this paper.

You will notice that you must include in your written program a plan for advising your outside contractors of the chemical hazards to which their employees may be exposed while they are working on your premises. It is equally -- if not more -- important that you make provisions for the opposite situation: Protecting your own employees from the chemical hazards brought onto your premises by outside contractors.

18. When Outside Contractors Work On Your Premises

Employers will -- at one time or another -- have outside contractors enter their premises to make repairs, do cleaning or painting, or perform some other function. Often that contractor will bring hazardous substances with him. You -- as the company in charge of the premises -- are responsible for communicating the hazards posed by those substances.

Suppose, for example, you hire a local painting contractor to strip and re-paint your office or factory walls. The liquid paint stripper he uses, however, contains a hazardous chemical and he uses it while your employees are working their regular jobs in the vicinity. Or he does his work at night while your employees are off duty -- but he leaves containers of his paint stripper on your premises during the day when your employees are working. That is _your_ responsibility. There must be an MSDS on that paint stripping material readily accessible to _your_ employees, _your_ employees must receive appropriate training thereon, and his containers must be labeled properly. If that is not done, the noncomplying employer is _you_ -- not the outside contractor.

These things becomes an even greater source of trouble when you have no idea what materials the outside contractor will be bringing on to your premises.

One way to handle this problem is to include appropriate provisions in the written contract you make with each of your contractors. Include specifications such as _when_ and _where_ the contractor will do his work and _what_ he is permitted -- or denied permission -- to bring onto your premises. You should include a requirement that, before his work commences, the contractor provide you with a complete list of what he will be bringing to your premises, an MSDS on everything he will use, and give you veto power over substances that will be harmful or are likely to cause you trouble. Another contractual provision you should insist upon is that your contractor observe the requirements of the hazard communication standard during the course of the contract. Then check up to make sure he does.

In some cases, you may want to require that your contractor provide training to those among your own employees who will be exposed to the hazardous substances the contractor brings onto your premises. There could be many other potential problems you will want to provide for in your contract. The important thing to remember is that the pre-printed "standard" contracts currently used by a number of business firms probably will not do. Nor should you permit an outside contractor to do work on your premises without a written contract. Even emergency plumbing or electrical repairs, obtained by consulting the yellow pages, could result in the introduction to your premises of substances regulated by the hazard communication standard.

Each contract for work to be done on your premises by an outside contractor should be tailored to your own particular situation. Moveover, **all** future such contracts should include hazard communication provisions, perhaps even indemnity clauses to protect you from liability caused by the contractor's failure to observe hazard communication requirements -- or by bringing a hazardous substance to your premises. These are matters that requires your careful attention and, in many cases, the services of legal counsel to guide you.

19. Written Hazard Evaluation Procedures

It was stated earlier that chemical manufacturers, importers and such, must evaluate all chemicals they handle and determine their hazard potential. This is part of the process for identifying which chemicals must have an MSDS.[15] The standard further provides, however, that the "procedures" to be followed in carrying out this responsibility must be committed to writing and made available upon request to employees, their union representatives, OSHA and NIOSH.[16] See subsection (d) of the standard.

Consequently, if you manufacture a product which in normal use or a foreseeable emergency might release, or otherwise result in exposure to, a hazardous chemical, you are required to make a determination of the potential hazard and you must make a written record of what you did in arriving at your determination. The processes for doing this are spelled out in subsection (d). They should be closely observed.

---

[15] That requirement is unique to the HCS. In many Right-to-Know laws, the chemicals upon which an MSDS are required are listed by name or by reference to published sources.

[16] You may want to carefully review the law that applies to disclosure of company records when you are considering a request for access to this document.

If you are required to furnish your customers with an MSDS, you should be cautioned against what some such manufacturers seem to be doing -- simply taking their supplier's MSDS and adapting it to their own product or obtaining a pro forms MSDS from a trade association. OSHA may contend that this practice overlooks the hazard evaluation requirement. A company that supplies its customers with a MSDS is obligated by law to conduct a hazard evaluation of the particular product involved and put in writing how it performed that hazard evaluation. If you rely upon one done by someone else, you may be relying upon something that was not done properly. That could be a source of trouble for you.

Even if you conclude that no MSDS is required for your product, you must be prepared to defend that judgment when asked why. Your written hazard determination is your answer. It is a matter that demands careful attention to detail.

The requirement that the procedures followed in making hazard evaluations be a written document almost certainly assures that, when some unforeseeable future event occurs, that document will be as closely examined as the sealed flight recorder after an airplane crash. You must be prepared for that prospect.

20. Trade Secret Protection

The manufacturers of many thousands of profit-making products are convinced that the reason for a particular product's success is the inclusion of that "extra ingredient" that no competitor's product contains. Colonel Sander's "secret recipe" for Kentucky Fried Chicken, for example. If forced to identify that ingredient, vital competitive advantage would be lost.

Government has not always been friendly to these trade secret claims. OSHA, for example, during the Carter Administration, adopted a "records access" regulation that requires disclosure of chemical identities of toxic substances regardless of any trade secret claims. See 29 C.F.R. § 1910.20(f). The Hazard Communication Standard, however, initially threw a friendly and protective arm around a manufacturer's trade secrets. Many state laws protect trade secrets from disclosure but, in this standard, OSHA went further than most states by defining the term much more broadly. However, as a result of a court challenge to this provision, OSHA has been forced to confine the trade secret protection to the limits imposed under state law. See subsection (i) of the standard.

No extensive discussion of trade secrets will be undertaken here because many employers who are subject to the standard have no trade secrets to protect. Those who do have trade secrets to protect should be aware of the fact that all specifically identifying information may be withheld from the MSDS, the Training Program and the Label itself where the chemical identity is a trade secret. The appropriate "hazard warning," however, must be given. Also, trade secrets must yield where there is a genuine medical emergency and must be disclosed to treating physicians, other health professionals and employee representatives but -- and this is a significant condition -- that disclosure can be conditioned in many cases upon a written assurance of confidentiality that can form the basis of a lawsuit -- and compensatory damages -- if not observed.

Many state and local right-to-know laws have completely different trade secret provisions. Employers with trade secrets to protect would be well advised to take the matter up with their attorney so they can be advised properly on these matters and fully prepared before any such emergency situation develops.

21. Preemption of Right-to-Know Laws

Since OSHA's inception, many industry representatives have regarded the agency as an adversary and have generally opposed nearly every new OSHA regulatory initiative. The HCS standard, however, resulted from a reversal of roles. Industry actually lobbied OSHA to adopt a hazard communication standard. It was not because either industry or OSHA had changed its spots. It was because many state and local governments were adopting "right-to-know" laws that industry did not like and they knew, or at least they hoped, that OSHA had the clout to nullify those laws by adopting a rule of its own.

OSHA's "clout" in that regard is the Federal Supremacy Clause of the U.S. Constitution. It provides, in effect, that when the Federal government occupies an area of the law, state jurisdiction over that area ceases. It is not quite that simple -- but that was what industry had in mind when it began its big push for OSHA to adopt "right-to-know" rules that would apply throughout the country.

There is no question that there has been a proliferation of state and local law in this field during the past few years. They cover different lists of substances that are deemed to be hazardous, contain different reporting requirements, serve different purposes, have different MSDS provisions, as well as different educational and training requirements. That is a rather burdensome situation for industry, especially for companies that do business in more than one state. The foregoing was not overlooked when OSHA, in the standard's preamble, listed its reasons for adoption of the hazard communication standard. Right up there among the chemically-related illnesses that inspired the standard was the following:

> "By promulgating a Federal standard, OSHA is in a position to reduce the regulatory burden posed by multiple State laws. In the final standard, OSHA preempts State law which deals with hazard communication requirements for employees in the manufacturing sector."

That statement was included in the original 1983 justification for the standard. Subsequent court rulings, however, held that the standard did not effectively preempt several Right-to-Know laws. See for example, Manufacturers Association of Tri-County v. Knepper, 801 F.2d 130 (3d Cir. 1986). OSHA therefore got much more specific when it expanded HCS coverage in 1987. It published the following statement at that time:

> The revised §1910.1200(a)(2) not only defines hazard communication as an "issue"

under the terms of the Act,[17] but also enumerates the generic areas addressed by the standard for purposes of establishing the parameters of preemption. Thus any State or local government provision requiring the preparation of material safety data sheets, labeling of chemicals and identification of their hazards, development of written hazard communication programs including lists of hazardous chemicals present in the workplace, and development and implementation of worker chemical hazard training for the primary purpose of assuring worker safety and health, would be preempted by the HCS unless it was established under the authority of an OSHA-approved State plan.

52 Fed. Reg. 31861, August 24, 1987.

Preemption, however, will frequently depend upon whether or not the state rule regulates the <u>workplace</u>. New Jersey, for example, requires employers to make environmental hazard surveys and supply them to the state and local health departments as well as local fire and police departments. Because those agencies are concerned with <u>public</u> health, the state requirements are not preempted by HCS which covers <u>occupational</u> health. <u>N.J. State Chamber of Commerce v. Hughey</u>, 774 F.2d 587 (3d Cir. 1985).

22. The Federal Right-to-Know Law

The Federal Government adopted its own right-to-know act in 1986 but it is neither administered nor enforced by OSHA. It is entitled the Emergency Planning and Community Right to Know Act of 1986, 42 U.S.C. §§ 11001 through 11050, and is administered and enforced by the Environmental Protection Agency (EPA). It

---

17/ The word "issue" has a special meaning under the Act. Under section 18(a), a state is <u>not</u> preempted from regulating "any occupational safety or health <u>issue</u> with respect to which <u>no</u> standard is in effect under section 6." 29 U.S.C. §667(a), emphasis added. Thus, the new OSHA language appears to <u>specifically</u> preempt state and local right-to-know laws by stating that HCS is an "issue" that Federal OSHA has covered by the adoption of the HCS.

includes emergency planning and notification requirements that first become operative at various dates in 1987 and 1988. Its provisions apply to all places of employment where substances determined by EPA to be extremely hazardous are present, as well as those companies required to prepare or have available an MSDS under the OSHA HCS.

The Federal right-to-know law is Title III of the Superfund Amendments and Reauthorization Act of 1986 ("SARA"). SARA revises and extends the authorities established under the Comprehensive Environmental Response, Compensation and Liability Act of 1980 ("CERCLA").

Although enacted as part of SARA, the Right-to-Know-Act is a free-standing provision of law and is thus separate from CERCLA. It was passed to create programs:

1. to provide the public with important information on the hazardous chemicals in their communities; and

2. to establish emergency planning and notification requirements which would protect the public in the event of a release of hazardous chemicals.

Both this Right-to-Know Act and the Hazard Communication Standard deal with transmittal of information regarding chemical hazards. The Hazard Communication Standard addresses the transmittal of information to workers about hazards of chemicals in the workplace. The Right-to-Know Act addresses transmittal of information to the public on the hazards of chemicals affecting the community.

The Federal Right-to-Know Act is organized into three subtitles:

1. Subtitle A establishes the framework for local emergency planning.

2. Subtitle B provides the mechanism for providing community awareness of hazardous chemicals in the locality. It includes requirements for:

a. the submission of material safety data sheets and emergency hazardous chemical inventory forms to state and local governments; and

b. the submission of toxic chemical release forms to the states and to the U.S. EPA.

3. Subtitle C contains general provisions concerning trade secret protection, enforcement, citizen suits, and availability of information to the public.

Two provisions in the new law, sections 311 and 312, mandate that employers who are required under the Occupational Safety and Health Act and regulations adopted under that Act to prepare or have available material safety data sheets for hazardous chemicals in their workplaces, must also submit chemical hazard information to State and local governments. Specifically, employers required by the OSHA HCS to create or maintain material safety data sheets for employees must also submit to the State emergency response commissions, the local emergency planning committee and the local fire department:

(1) A material safety data sheet for each hazardous chemical for which a data sheet is available (section 311); and

(2) An emergency and hazardous chemical inventory form (section 312).

The public may request material safety data sheets and inventory information from the local planning committee.

Because all employers are currently subject to the OSHA HCS and required to create or maintain data sheets for the hazardous chemicals present in their workplaces, they must also comply with the community reporting requirements of the Emergency Planning and Community Right-to-Know Act.

EPA has established a toll-free hotline to answer questions concerning the requirements: Chemical Emergency Preparedness Program Hotline, 1-800/535-0202; in Washington, D.C. at 1-202/479-2449.

Additional implementing rules and regulations are currently being developed. There is also additional legislation of a similar nature under consideration at the federal, state and local levels. Employers must therefore keep up to date in order to know what's new in right-to-know.

23. Civil Liability Aspects

The HCS does not create any new cause of action, but it does require the development and maintenance of information. It is that information that could turn out to be a source of trouble for an employer in a civil liability proceeding because those information requirements will make it much easier for future litigants to prove their cases. The "smoking gun" that could be used to establish the liability of the employer-defendant in claims brought by one or many claimants seeking recovery for chemically-related diseases could be the MSDS, the hazard communication program, the hazard determination or even the container label.

All of these "communications" have one thing in common: They are in writing and, in the eyes of the law, they are <u>documents</u>. Few people need to be reminded of the power of documentary evidence. It has literally brought down emperors, kings, chiefs of state, financial wizards and business empires, not to mention those violent criminals who somehow managed to silence all eyewitnesses to their crimes but got caught by ubiquitious IRS forms. What may seem a document entirely innocent in its implications may in fact prove, in certain circumstances, to be dynamite. Former president Richard Nixon discovered that.

Pick up an MSDS for example. Included among the information it must contain is the health hazards of the listed chemical including any medical conditions that are generally recognized to be aggravated by exposure to it. If one such condition was pregnancy, for example, but pregnant employees nevertheless worked around the chemical, it is not hard to imagine the use of that MSDS as evidence against the employer in a liability suit resulting from terminated pregnancies or deformed babies.[18]

The employer could defend against a claim of that kind on the grounds that he didn't know of the pregnancies, that the concentration of the chemical that was used in his business was fare less than what is needed to cause harm, or a number of other grounds - and he might win. But he will have to contend with the MSDS, a document that he may not have had to face before HCS and Right-To-Know made its creation, distribution and retention a requirement of law. The very _existence_ of the documents required under HCS and Right-To-Know will serve as the basis for claims and lawsuits that otherwise might never have been considered.

24. Observing the Written Precautions

Most MSDSs contain warnings that are often unobserved such as "use only with adequate ventilation," "may aggravate some respiratory conditions," and the like. Some state that employees with pre-existing medical conditions shouldn't work around the product. Regrettably, some employers pay little heed to these warnings. Some don't even read them. They do so at their peril.

Companies who use chemicals frequently claim that the precautions and warnings that chemical manufacturers include in

---

[18] That kind of problem appears to be obtaining greater recognition in recent years. In its December 9, 1988 edition, the Washington _Post_ printed a lengthy article on its front page about an unusually high number of reported miscarriages among women working in the newsrooms of the _USA Today_ newspaper in Arlington, Virginia. Included in the article was a box entitled "Clusters of Pregnancy Problems Among VDT Users," the source of which was listed as: New York State Department of Health. That box contained a list of twelve workplaces where miscarriages, severe defects and other abnormalities have been reported in unusually numbers in the 1980s. The article also reported that NIOSH has been studying the matter. No mention of any MSDS precautions was made in the article nor of any civil liability suits against employers. Nevertheless, the potential is there and employers must concern themselves with it.

the MSDSs are overstated. They may be right because some chemical manufacturers do indeed try to insulate themselves from liability by providing for every conceivable precaution when using their product.

The users, however, should _never_ ignore the manufacturer's MSDS warnings and requirements. That practice will come back to haunt them when the failure to require observance with the MSDS is used as the basis for a multi-million dollar lawsuit against the company -- or a criminal prosecution against a company official. The jury in such a case is unlikely to treat very kindly an employer-defendant's rejoinder that he hadn't read the MSDS and didn't know about the manufacturer's MSDS warnings. A judge or jury would be even less receptive to an employer's claim that he decided on his own to ignore written MSDS warnings or instructions because he felt that were unnecessary and overstated.

Then there are other HCS documents to contend with. The "hazard determination" provision of the HCS, 29 C.F.R. §1910.1200(d), requires chemical manufacturers and importers to evaluate chemicals, determine if they are hazardous, and commit to writing the procedures used in so doing. Surely that document will find its way into future liability suits whenever, upon 20/20 hindsight, it appears that a chemical's actual hazards were not properly assessed at the time the determination was made.

Years after it was created, that document is certain to be examined with a fine tooth comb after disease or disaster has struck someone down. And that process might establish that faulty procedures were followed in making the hazard determination. Will that be enough to prove liability? Your ess is as good as anyone's. But, once again, the existence of such a document will provide a basis for future claims. Knowledge of that prospect should also provide an incentive for the employer to devote very careful attention to the requirement.

The foregoing is simply an illustration of how HCS and Right-to-Know documents can be used. Similar problems can arise from the container labels and the employer's hazard communication program, both of which must also be written documents.

Because the required existence of those documents is of very recent origin, court decisions on their use in civil liability cases have not yet been widely reported. The potential, however, is a very real one.

25. What Employers Must Do

Employers should derive four mandates from the foregoing:

(1) Know both the OSHA HCS and the EPA right-to-know requirements and observe them.

(2) Learn and observe the applicable state and local right-to-know rules of those places where you do business and keep abreast of current developments at all levels: Federal, state and local;

(3) Carefully analyze every MSDS on every product that is on your premises. Make certain that you are observing every MSDS precaution and warning. If you are not or cannot, get rid of the product at once.

(4) Be aware that the many things that are put in writing pursuant to those requirements may be used against you, not only by government agents, but by private individuals, community, labor or environmental groups, or even by other employers.

The case law in the HCS/Right-to-Know area of the law is in its infancy. So, too are the administrative interpretations.[19/] They will surely have an impact on what has been discussed here but that impact cannot now be predicted.

19/ OSHA has its own lengthy interpretation of HCS, which is designated as OSHA Instruction CPL 2-2.38. It was first issued on May 16, 1986 and has since been revised several times. The most recent edition of that Instruction was issued on October 22, 1990. It is informative but not legally binding because it has not been published in the Federal Register.

APPENDIX C:
OSHA INSTRUCTION CPL 2-2.38C INSPECTION PROCEDURES FOR THE HAZARD COMMUNICATION STANDARD, 29 CFR 1910.1200, 1915.99, 1917.28, 1918.90, 1926.59, AND 1928.21

Subject: Inspection Procedures for the Hazard Communication Standard, 29 CFR 1910.1200, 1915.99, 1917.28 1918.90, 1926.59, and 1928.21

A. <u>Purpose.</u> This instruction establishes policies and provides clarifications to ensure uniform enforcement of the Hazard Communication Standard (HCS).

B. <u>Scope.</u> This instruction applies OSHA-wide.

C. <u>References</u>.

   1. OSHA Instruction CPL 2.45B, June 15, 1989, the Revised Field Operations Manual (FOM).

   2. OSHA Instruction STP 2-1.117, August 31, 1984.

   3. Voluntary Training Guidelines, Vol. 49, FR 30290, July 27, 1984.

   4. 29 CFR 1910.20, Access to Employee Exposure and Medical Records.

   5. 29 CFR 1910.1047, Ethylene Oxide.

   6. 29 CFR 1910.1000, Air Contaminants, Vol. 54, FR 2332, January 19. 1989.

   7. The HCS was recodified and referenced as 29 CFR 1910.1200 for General Industry, 1915.99 for Shipyard Employment, 1917.28 for Marine Terminals, 1918.90 for Longshoring, 1926.59 for Construction and 1928.21 for Agriculture. For convenience this instruction will reference only applicable paragraphs. The appropriate sections of the CFR shall be referenced for citation purposes when inspections are performed in those respective industries.

D. <u>Cancellation.</u> OSHA Instruction CPL 2-2.38B, August 15, 1988, is canceled.

E. <u>Action</u>. OSHA Regional Administrators and Area Directors shall use the guidelines in this instruction to ensure uniform enforcement of the HCS. The Directorate of Compliance Programs will provide support as necessary to assist the Regional Administrators and Area Directors in enforcing the HCS.

F. **Federal Program Change**. This instruction describes a Federal Program change which affects State programs. Each Regional Administrator shall:

1. Ensure that this change is forwarded to each State designee.

2. Explain the technical content of the change to the State designee as requested.

3. Advise the State designees that as a result of further court actions, all provisions of the Federal HCS are now in effect in all segments of industry. The compliance date for programmed inspections in the construction industry and the three previously disapproved provisions was extended to March 17, 1989. States not already enforcing in all industries were expected to have done so by that date.

4. Ensure that State designees are asked to acknowledge receipt of this Federal program change in writing to the Regional Administrator as soon as the State's intention is known, but not later than 70 calendar days after the date of issuance (10 days for mailing and 60 days for response). This acknowledgment must include the State's intention to follow OSHA's policies and procedures described in this instruction, or a description of the State's alternative policy and/or procedure which is "at least as effective" as the Federal policy and/or procedure or of the reasons why the change should not apply to that State.

5. Ensure that the State designees submit a plan supplement, in accordance with OSHA Instruction STP 2.22A, Ch-3, as appropriate, following the established schedule that is agreed upon by the State and Regional Administrator to submit non-Field Operations Manual/Technical Manual Federal Program Changes.

    a. If a State intends to follow the revised inspection procedures described in this instruction, the State must submit either a revised version of this instruction, adapted as appropriate to reference State law, regulations and administrative structure, or a cover sheet describing how references in this instruction correspond to the State's structure. The State's acknowledgment letter may fulfill the plan supplement requirement if the appropriate documentation is provided.

    b. If the State adopts an alternative to Federal enforcement inspection procedures, the State's plan supplement must identify and provide a rationale for all substantial differences from Federal procedures in order for OSHA to judge whether a different State procedure is as effective as a comparable procedure. An alternative enforcement policy would presumably be necessary in a State with a right-to-know law or a different hazard communication standard.

    c. Any State which has a right-to-know law shall also document in the plan supplement how enforcement of the right-to-know law substitutes for, relates to or interfaces with the hazard communication standard, and how the State maintains separation of any public/community right-to-know enforcement activities from its approved State plan workplace operations.

6. After Regional review of the State plan supplement and resolution of any comments thereon, forward the State submission to the National Office in accordance with established procedures. The Regional Administrator shall provide a judgment on the relative effectiveness of each substantial difference in the State plan change and an overall assessment thereon with a recommendation for approval or disapproval by the Assistant Secretary.

7. Review policies, instructions and guidelines issued by the State to determine that this change has been communicated to State personnel.

G. **Special Identifiers**. The sections of this instruction which are marked with an asterisk (*) have particular relevance to construction employers.

H. **Background**. A final Hazard Communication Standard   *
(HCS), 29 CFR 1910.1200, covering the manufacturing sector, Standard Industrial Classification Codes (SIC) 20-39, was published in the Federal Register on November 25,

1983 (48 FR 53280). As a result of a court challenge, OSHA was ordered by the U.S. Court of Appeals for the Third Circuit to expand the scope of the standard without further rulemaking.

1. On August 24, 1987, a final rule covering all employers was published in the Federal Register. Due to subsequent court and administrative actions, OSHA was prevented from enforcing the rule in the construction industry and from enforcing in all industries three requirements dealing with providing and maintaining material safety data sheets (MSDSs) on multi-employer worksites, coverage of consumer products, and the coverage of drugs in the nonmanufacturing sector.

2. As a result of the February 21, 1990, Supreme Court decision (see <u>Dole, Secretary of Labor, et. al., v. United Steelworkers of America et. al.</u>, No. 88-1434), all provisions of the rule are now in effect for all industrial segments, including the three previously stayed provisions mentioned above. OSHA extended the compliance date until March 17, 1989, for programmed inspections in the construction industry.

I. <u>Organization of this Instruction</u>. Compliance guidelines are addressed within the main part of this instruction. Clarifications, interpretations, review aids and other information are provided in Appendices A through D. This format will permit easier updating and additions, as enforcement experience provides more information regarding these areas.

1. Appendix A of this instruction provides clarifications of provisions of the standard where significant interpretations have been necessary to ensure uniform enforcement and understanding.

2. Appendix B provides a sample letter for inquiries regarding missing or deficient MSDS and labels.

3. Appendix C provides general guidelines for evaluation of hazards.

4. Appendix D provides a guide for reviewing MSDS.

J. **Inspection Resources.** Compliance safety and health officers (CSHOs) shall evaluate employer compliance with the HCS during the course of <u>all</u> inspections. (See the FOM, Chapter III, D.7.a.2.)

1. Both safety and health CSHOs shall evaluate employer compliance with the written program requirements, use of labels, availability of MSDS and appropriate training.

2. CSHOs of one discipline shall consult with those of the other when specific expertise is necessary to evaluate elements of the employer's program.

K. **Inspection Guidelines.** The following guidelines apply to all inspections conducted to determine compliance with the HCS:

1. **Inspection Guidance.** The HCS incorporates both specification and performance requirements which are result-oriented, thereby providing goals for achievement and allowing employers the flexibility to develop a program suitable for their particular facility. In evaluating compliance with the rule, CSHOs should always consider whether the intent of the provisions have been met. CSHOs must exercise a high level of professional judgment during compliance inspections. The standard itself, and the preamble accompanying it, are to be consulted for further guidance.

2. **Special Documentation.** In addition to those items required by the FOM, Chapter IV, C. 8. as applicable, when citations are recommended, the CSHO shall document the following on the OSHA-1B or, as appropriate, elsewhere in the case file:

   a. Name of the chemical(s).

   b. Name of the person preparing the hazard determination, written program, label, MSDS, etc. and for whom they work.

   c. CSHOs shall ensure that the number of employees who may be exposed (including potential exposure) to the chemical in the establishment is documented.

d. If a chemical manufacturer, importer, or distributor is inspected, indicate the name of a downstream employer who receives the chemical, including company name, address, and potential or actual downstream employee exposure.

e. Health and physical hazards.

f. If practical, include a photocopy or a photograph of inaccurate and/or any incomplete label(s)/MSDS in the case file. Otherwise document the specific deficiency in the case file. If the volume of inaccurate/incomplete MSDS cannot reasonably be included in the file, then a representative number should be documented, indexing those referenced in the citation.

3. <u>Scope and Application - Paragraph (b)</u>. The HCS requires labels and MSDS to be transmitted from chemical manufacturers and importers to distributors to employers to employees. No barrier to this information flow is permitted.

   a. This paragraph outlines exemptions to full coverage of the standard. A complete exemption from all requirements of the HCS applies for only those items listed under (b)(6) and should not be confused with the <u>labeling</u> exemptions at (b)(5) which only apply when chemicals are subject to the labeling requirements of certain Federal agencies.

   b. Laboratories and sealed containers are dealt with in a limited fashion as per paragraphs (b)(3) and (b)(4).

   c. <u>Inspection Guidelines</u>. As explained in H.2. of this instruction, the HCS has been fully enforceable in all SIC's since March 17, 1989. *

      (1) The Scope and Application paragraph (b) of the HCS requires "<u>all</u> employers to provide information to their employees about the hazardous chemicals <u>to which they are exposed</u>, by means of a hazard communication

program, labels and other forms of warning, material safety data sheets and information and training." (Emphasis added.)

(2) The expansion of the standard to all industries via the August 24, 1987, final rule obligates all employers to comply with the provisions of the HCS. Employers must provide their employees with information on hazardous substances which are known to be present at the worksite.

(3) The scope paragraph clearly states that the HCS applies to employers if they know hazardous chemicals are present in a manner that employees may be exposed, regardless of whether the employer has created the chemical exposure. The multi-employer worksite provisions of paragraph (e)(2) ensure that employers are able to obtain the information they need to be able to meet these obligations.

(4) In some cases, a hazardous chemical may be present for a long period of time without an employee exposure until repair or demolition activities are performed. By way of example, employers involved in work operations where jackhammers are being used to break up a sidewalk know that they are exposing their employees to a hazardous chemical (silica), even though they did not "bring" the hazard to the site. Even though other provisions of the standard may not be enforceable (MSDS and labels), the employer should still develop a hazard communication program to inform their employees "about the hazardous chemicals to which they are exposed." Employers may utilize their already existing hazard communication program to communicate information on these types of hazards to their employees, as per paragraph (e)(3).

4. <u>Hazard Determination - Paragraph (d)</u>. Only chemical manufacturers and importers are required to perform hazard determinations on all chemicals they produce or

import, although distributors and employers may choose to do so. Hazard determination procedures must be in writing and made available, upon request, to employees, the National Institute for Occupational Safety and Health (NIOSH), and OSHA. Appendix C is provided as a guide for use when assessing the hazard evaluation procedures.

a. <u>Inspection Guidelines</u>. The adequacy of a company's hazard determination program can be assessed primarily by examining (or reviewing) the outcome of that determination; i.e., the accuracy and adequacy of the information on labels and MSDS. The written hazard evaluation procedures generally describe the process followed; they do not have to address each chemical evaluated. The chemical manufacturer, importer, employer or distributor performing the hazard determination ("the preparer"), shall be asked to forward the written hazard determination procedures to the Area Director when they are not immediately available at the establishment. A reasonable time period, not exceeding 5 working days, shall be allowed for receipt in the Area Office.

(1) Although not required, many companies will keep records of individual chemical evaluations. In the event of a finding by the CSHO of an inaccurate determination, as indicated by inaccurate information on the MSDS or label, these records may be useful in identifying where the company's evaluation differed from OSHA's, and for documentation of appropriate violations.

(2) In general, the hazard evaluation procedures should address the following:

(a) The sources of information to be consulted. Evaluators should have access to a wide range of sources. While well-known chemicals could be adequately evaluated by consulting established reference texts, others will require searches of bibliographic data bases.

(b) Criteria to be used to evaluate the studies, including those parameters addressed by the HCS (i.e., statistical significance; whether or not the evaluation was conducted according to established scientific principles).

(c) A plan for reviewing information to update the MSDS if new and significant health information is found.

(3) The hazard evaluation must include an assessment of both physical **and** health hazards. The chemical manufacturer or importer must consider the potential exposures that may occur when downstream employers use the product, and address the hazards that may result from that use on the labels and MSDS prepared for the product. It is important to note that employee "exposure" as defined by the HCS includes any route of entry (inhalation, ingestion, skin contact or absorption) and also includes potential (e.g., accidental or possible) exposure, including foreseeable emergencies. Only by considering all these factors can the chemical manufacturer or importer truly assess the hazards encountered during anticipated use of his product. The mere presence of a chemical in a product does not necessarily result in coverage; it must be available for exposure.

(4) Evaluations with respect to carcinogen labeling and MSDS notations are addressed in those respective sections below as well as in Appendix A which also contains specific information on mineral oils.

b. <u>Citation Guidelines</u>. Citations for violations of paragraph (d)(1) shall be issued when the preparer has failed to perform a hazard determination. Paragraphs (d)(2), (d)(3) and (d)(4) of the standard shall be used, as appropriate.

(1) If the preparer has developed MSDS but does not have the written procedures available that were used to determine the hazards of the chemical(s), then a violation of paragraph (d)(6) exists and shall be recommended for citation.

(2) If the preparer has not developed an MSDS and no written procedures are available, then apparent violations of both paragraphs (d)(1) and (d)(6) exist and shall be recommended for citation. (Refer to K.7.b. of this instruction for guidance.)

(3) Chemical manufacturers or importers are not required to test their products to evaluate their hazards. If a mixture has been tested, the resulting data would apply. If it has not been tested as a whole, the mixture is assumed to present the same hazards as its component parts. If the employer chooses to rely on upstream chemical manufacturers' hazard determinations for the component parts of his mixture, he may do so but must so specify in his written hazard determination procedures. MSDS for each of the component parts must be physically grouped together in order to meet the chemical manufacturer's hazard determination requirements. Certain information has to be provided for the mixture as a whole for the combined MSDS; e.g., identity, manufacturer's name, address, etc.

5. **Written Hazard Communication Program, Paragraph (e)**. * CSHOs shall review the employer's written hazard communication program to determine if all applicable requirements of paragraph (e) have been addressed. The HCS obligates all employers who may expose their employees to hazardous chemicals to develop a written program, regardless of whether or not they introduced the hazard into the workplace.

   a. **Inspection Guidelines**. Ideally, and if readily available, the written program should be reviewed first, prior to ascertaining whether the elements of the program have been implemented in the workplace.

(1) The CSHO shall determine whether or not the employer has addressed the issues in sufficient detail to ensure that a comprehensive approach to hazard communication has been developed.

(2) In general, the written program should consider the following elements where applicable:

   (a) <u>Labels and Other Forms of Warning</u>.

      <u>1</u>   Designation of person(s) responsible for ensuring labeling of in-plant containers.

      <u>2</u>   Designation of person(s) responsible for ensuring labeling on shipped containers.

      <u>3</u>   Description of labeling system(s) used.

      <u>4</u>   Description of written alternatives to labeling of in-plant containers, where applicable.

      <u>5</u>   Procedures to review and update label information when necessary.

   (b) <u>Material Safety Data Sheets</u>.

      <u>1</u>   Designation of person(s) responsible for obtaining/maintaining the MSDS.

      <u>2</u>   How such sheets are to be maintained (e.g., in notebooks in the work area(s), via a computer terminal, in a pick-up truck at the jobsite, via telefax) and how employees obtain access to them.

<u>3</u> Procedure to follow when the MSDS is not received at the time of the first shipment.

<u>4</u> For chemical manufacturers or importers, procedures for updating the MSDS when new and significant health information is found.

(c) <u>Training</u>.

<u>1</u> Designation of person(s) responsible for conducting training.

<u>2</u> Format of the program to be used (audiovisuals, classroom instruction, etc.).

<u>3</u> Elements of the training program--compare to the elements required by the HCS (paragraph (h)).

<u>4</u> Procedures to train new employees at the time of their initial assignment and to train employees when a new hazard is introduced into the workplace.

<u>5</u> Procedures to train employees of new hazards they may be exposed to when working on or near another employer's worksite (i.e., hazards introduced by other employees).

<u>6</u> Guidelines on training programs prepared by the Office of Training and Education entitled "Voluntary Training Guidelines" (Vol. 49 <u>FR</u> 30290, July 27, 1984) can be used to provide general information on what constitutes a good training program.

(d) <u>Additional Topics To Be Reviewed</u>.

<u>1</u>   Does a list of the hazardous chemicals exist and if so, is it compiled for each work area or for the entire worksite and kept in a central location?

<u>2</u>   Are methods the employer will use to inform employees of the hazards of <u>non-routine</u> tasks outlined?

<u>3</u>   Are employees informed of the hazards associated with chemicals contained in unlabeled pipes in their work areas?

<u>4</u>   Does the plan include the methods the employer will use at multi-employer worksites to inform other employers of any precautionary measures that need to be taken to protect their employees?

<u>5</u>   For multi-employer workplaces, are the methods the employer will use to inform the other employer(s) of the labeling system used described?

<u>6</u>   Is the written program made available to employees and their designated representatives?

b. <u>Citation Guidelines</u>. Generally, all violations of paragraph (e) shall be grouped with the violated element(s) listed in the subparagraphs of (e) and/or violations of paragraphs (f), (g) and (h) as appropriate, since (e)(1) is the only provision under paragraph (e) which addresses the development, implementation and maintenance of the written hazard communication program. Specific citation guidance is given below:

(1) Paragraph (e)(1) shall be cited by itself when no program exists (i.e., when no program has been developed). Paragraph (e)(1) shall

also be cited in instances where the written program is not maintained at a fixed worksite location. For certain mobile or multi-employer worksite situations, see guidance given in Appendix A, Section (e)(2), discussion beginning on page A-15.

(2) When an employer's written program exists but is found to be deficient (i.e., has not been implemented as witnessed by the inadequacies of the other requirements of the standard), paragraph (e)(1) shall be cited and grouped as separate violations with separate penalties with the elements of the standard required in subparagraphs of (e) and/or paragraphs of (f), (g), and/or (h). An example follows: An employer has developed a written program but it has not been implemented in the workplace--no training has been provided and MSDSs are not available to employees. In this situation two separate violation items shall be recommended for citation: (e)(1) grouped with (h) as a separate violation and penalty and (e)(1) grouped with (g)(8) as a second violation with separate appropriate penalty.

(3) Paragraph (e)(1) shall also be cited when an employer has not developed a written program and yet is exposing his employees to chemical hazards which are known to be present in the workplace and which are created by another employer.

(4) OSHA's compliance and enforcement policies for multi-employer worksites are set forth in the FOM, Chapter V, Sections F.1 and 2., which state that with regard to working conditions where employees of more than one employer are exposed to a hazard, the employers "with the responsibility for creating and/or correcting the hazard" shall be cited for violations of OSHA standards that occur on a multi-employer worksite. In these situations, normally citations for violations shall be issued to each of the

exposing employers <u>as well as to</u> the employer responsible for correcting or ensuring the correction of the condition (which is usually the controlling employer or general contractor).

6. <u>Labels and Other Forms of Warning, Paragraph (f)</u>. * Labels or other markings on each container of chemicals must include the identity and appropriate hazard warnings. Labels on shipped containers must also include the name and address of the chemical manufacturer, importer, or other responsible party.

    a. <u>Inspection Guidelines.</u> CSHOs shall determine that containers are labeled, that the labels are legible, and that the labels are prominently displayed.

        (1) Labels must be in English. Labels and MSDS may also be printed in additional languages.

        (2) The accuracy of the label information is to be assessed for a representative number of chemicals. The CSHO shall determine whether the label identity can be cross-referenced with the MSDS and the list of hazardous chemicals.

        (3) CSHOs must consider alternate labeling provisions (for example tags or markings) for containers which are too small to accommodate a legible label.

        (4) CSHOs shall evaluate the effectiveness of in-plant labeling systems through a review of the employer's training program and MSDS procedures. Such evaluation shall include interviews with employees to determine their familiarity with the <u>hazards</u> associated with chemicals in their workplace. An effective program is one that ensures that employees are aware of the hazardous effects (including target organ effects) of the chemicals to which they are potentially exposed.

(5) Guidelines for referrals regarding inadequate labels are dealt with in this instruction at K.7.a.(7) and (8).

b. <u>Citation Guidelines.</u> Chemical manufacturers shall be cited for appropriate paragraphs (f)(1)(i) through (f)(1)(iii) of the standard when deficiencies are found relating to products that are shipped downstream. Paragraphs (f)(5)(i) and (f)(5)(ii) of the standard shall be cited when a hazardous chemical is created and/or used only in-house. (See also K.7.b.)

7. <u>Material Safety Data Sheets, Paragraph (g)</u>. The *
standard requires chemical manufacturers and importers to develop or obtain a material safety data sheet for each hazardous chemical they produce or import.

   a. <u>Inspection Guidelines</u>. Distributors and employers may, at their option, develop MSDSs. CSHOs should inform them as well as chemical manufacturers and importers that the Material Safety Data Sheet, OSHA Form 174, is available for this purpose. The CSHO shall evaluate the compliance status of this provision by examining a sample of MSDSs to determine that the MSDSs has been obtained or developed and prepared in accordance with the requirements of paragraphs (g)(2)-(5) of the standard and to ensure that the information regarding the health and physical hazards is technically accurate. If MSDSs are not updated when new information becomes available, the hazard determination performed by the chemical manufacturer or importer is deficient.

      (1) The number of MSDSs and the particular MSDS selected for review will depend upon several factors, such as:

         (a) The number of chemicals in the workplace.

         (b) The severity of the hazards involved.

         (c) The completeness of the MSDS in general.

(d) The volume of the chemicals used.

(2) The CSHO is to complete this review by following the procedures outlined in <u>Hazard Evaluation Procedures,</u> Appendix C of this instruction. The CSHO shall also use available literature and computer references in the Area Office as well as Appendix D, <u>Guide to Reviewing MSDS Completeness</u>, in reviewing MSDS.

NOTE: Published MSDS reference files are copyrighted, and, therefore, must NOT be copied for distribution to the public.

(3) In addition, each Area Office has access to physical and health hazard data on the OSHA Computerized Information System (OCIS). If the hazard information is not available or cannot be obtained in the Area Office, then the Regional Office shall be consulted. If the Regional Office does not have information on the chemical in question, then the Regional Office shall contact the Technical Data Center.

(4) Published MSDSs, if used, are a screening resource for the CSHO. The information on these MSDSs has not been evaluated by OSHA to determine if it is accurate or required in every situation. They should be used to help identify which areas require further research or where information is lacking on the MSDS being reviewed.

(5) The following items shall be considered when reviewing the MSDSs:

(a) Do employers have an MSDS for each hazardous chemical used?

(b) Does each MSDS contain information which adequately addresses at least the 12 elements required by the standard at (g)(2)(i)-(xii)?

(c) Are all sections of the MSDS accurately completed?

(6) The CSHO shall ensure compliance with the MSDS transmission provisions of the standard by reviewing the chemical manufacturer's, importer's, or distributor's program for transmitting the MSDSs and updated MSDSs to downstream customers.

(7) <u>Referral Procedures Where an Employer's MSDS/Label is Inadequate or Deficient</u>. Where employers are relying on the MSDS/label supplied by chemical manufacturers or importers, the following procedures apply:

(a) Employers are not to be held responsible for inaccurate information on the MSDS/label which they did not prepare and they have accepted in good faith from the chemical manufacturer, importer or distributor.

(b) The CSHO shall take copies of the MSDS/label with inaccurate information back to the Area Office for referral to the appropriate State Plan State or Area Office. Before making the referral, the Area Director shall write to the supplier requesting action in 30 days or less using the sample letter in Appendix B of this instruction. As an option, the Area Office may call the supplier, but if a prompt response is not received, a letter shall be sent. This may be done even if the supplier is outside the jurisdictional area of the Area Office.

(c) If the manufacturer or supplier fails to respond within a reasonable time, a referral (OSHA-90 Form), with complete background information attached, shall be sent to the State Plan State or Area Office within whose jurisdiction the supplier or manufacturer does business.

(d) The Area Office within whose jurisdiction the upstream supplier or manufacturer is located shall then ensure that an abbreviated (HCS) inspection is conducted or that a letter is written in accordance with the referral procedures in the FOM, Chapter IX, B.3.b.

(e) The findings and the MSDS(s) and/or labels obtained shall be sent to the referring office.

(f) The Regional Administrator shall coordinate with State designees to ensure that referrals from State plans are handled in similar manner. OSHA will not act on a referral from a State if it is for the purpose of obtaining an MSDS for inclusion in a State-maintained MSDS file and/or repository.

(8) **Referral Procedures for Distributors.** When a distributor has not received an MSDS from the supplier, the CSHO shall recommend that the distributor write to the chemical manufacturer, and, if applicable, other distributor who supplied the chemical. If at the end of the abatement period, the distributor has failed to receive the MSDS, the Area Director shall follow the referral procedures outlined in K.7.a.(7)(b) through (f) of this instruction.

b. **Citation Guidelines.** Citations shall be issued to the employer only when MSDS/labels are missing.

(1) If MSDS/labels are missing or have not been received for a hazardous chemical(s), the employer shall be cited unless a good faith effort has been made to obtain the information.

(a) A copy of a letter or documentation of a phone call to the supplier are examples of methods for establishing a good faith effort. An employer contacting OSHA for assistance in obtaining the missing information is also an excellent example of a good faith effort.

(b) Area Offices should expect to receive requests from employers to assist them in obtaining MSDSs or labels in situations when an inspection has not been conducted. If the Area Director determines that the employer has tried to obtain the information, and has not been able to do so, a letter and/or telephone call from the Area Office to the supplier or manufacturer is the appropriate action in this situation as well.

(c) If a citation will be issued to the employer for lack of a MSDS/label, where the employer has failed to document that a good faith effort has been made to obtain them, CSHOs shall recommend that the employer write to both the direct supplier and to the manufacturer for the MSDS or label.

    1   CSHOs shall inform employers that it is their responsibility to contact OSHA before the expiration of the abatement date to request a petition to modify abatement or else be subject to a failure to abate if abatement is not accomplished. If at the end of the abatement period the employer still has failed to receive the requested information, the Area Director shall call and/or send a certified letter to the manufacturer, importer, or distributor to obtain the required information. (See sample letter in Appendix B.)

    2   If the distributor failed to transmit the MSDS to the employer,

the distributor shall be cited for violation of paragraph (g)(7) of the standard with a short abatement date unless the distributor did not receive the MSDS from the chemical manufacturer, importer, or distributor. In such cases the abatement period will generally be 30 days.

(2) Any party who changes the label/MSDS (for example, changing the name or identity of the chemical) then becomes the responsible party for the change regardless of whether they are a chemical manufacturer, distributor or user employer. In cases where a distributor _adds_ its name to the MSDS and those MSDSs are inaccurate or incomplete, citations shall _not_ be issued to the distributor. Distributors, however, who _substitute_ their names on the MSDS or change it in any way become the "responsible party" and must be able to supply the required additional information on the hazardous chemical and appropriate emergency procedures, if necessary. Failure to be able to provide the additional information will result in a violation of (g)(2)(xii) of the standard if noted upon inspection.

(3) On multi-employer worksites, citations for violations of (g)(8) of the standard shall be issued to the employer responsible for providing or making the MSDS(s) available, as discussed below. A citation for violation of (e)(2) of the standard shall concurrently be issued in any of the instances listed where there is evidence that an employer has failed to effectively implement and enforce its hazard communication program.

   (a) If an employer on a multi-employer worksite brings hazardous chemicals onto that site and fails to inform other employers about the presence of those chemicals and/or the availability of the

MSDS(s), that employer shall be cited for violation of (g)(8) grouped with (e)(2).

 (b) <u>Central Location</u>. If the employer's method to provide other employers with MSDS(s) involves the use of a central location, and the MSDS(s) is not available at that location, then the employer shall be cited for violation of (g)(8).

 (c) <u>Controlling Employer</u>. If the employer's method involves using a general contractor or other employer as an intermediary for storage of the MSDS(s), and that intermediary employer has agreed to hold and provide ready access to the MSDS(s), then that other employer becomes the <u>controlling employer</u>, who is then responsible for ensuring the availability of the MSDS(s).

  <u>1</u> The controlling employer (e.g., general contractor) shall therefore normally be cited for violation of (g)(8) if the MSDS(s) is not available; however:

  <u>2</u> If the MSDS(s) is not available because the subcontractor failed to provide it, then the subcontractor shall instead be cited for violation of (g)(8).

(4) The FOM discusses penalty factors for shipped containers at Chapter IV, Section C.8.

8. <u>Employee Information and Training, Paragraph (h)</u>. *

 a. <u>Inspection Guidelines.</u> The training requirements of the HCS will generally complement rather than satisfy the existing training requirements contained within other OSHA standards (i.e., expanded health standards, construction requirements, etc.).

(1) CSHOs shall continue to ensure that employers' obligations under specific training provisions of other standards are met. There will also be instances where there is an overlap in the training requirements of 29 CFR 1926.21, Safety Training and Education, and the HCS. In those instances where the training deficiency is covered by both standards (1926.21 and 1926.59) the CSHO shall issue a citation for 29 CFR 1926.59, which is the more specific standard.

(2) Training programs must be evaluated through program review and discussion with management and employees. All elements of training and information stated in the standard must be addressed. The following additional questions provide a general outline of topics to be reviewed:

  (a) Has a training and information program been established for employees exposed to hazardous chemicals?

  (b) Is this training provided at the time of initial assignment and whenever a new hazard is introduced into work areas?

  (c) Have all new employees at this location received training equivalent to the required initial assignment training?

  (d) Was training subject matter organized by:

   1 Specific chemical?

   2 Categories of hazard?

(3) The Voluntary Training Guidelines (Vol. 49 FR 30290, July 27, 1984) may also be helpful in assessing the effectiveness of the employer's training program.

(4) Employee interviews will provide general information to the CSHO regarding the training program. Obviously, it cannot be expected that employees will totally recall all information and be able to repeat it. Employees must be aware of what hazards they are exposed to, know how to obtain and use information on labels and MSDS, and know and follow appropriate work practices. However, if the CSHO detects a trend in employee responses that indicates training is not being conducted, or is conducted in a cursory fashion that does not meet the intent of the standard, a closer review of the written program and its implementation may be necessary. The purpose of the standard is to reduce chemical source illnesses and injuries through the transmission hazard information. This can occur only if employees receive the information in usable form through appropriate training.

(5) Paragraph (h) requires that information and training be provided to employees regarding the hazards of all chemicals in their work areas including by-products and hazardous chemicals introduced by another employer, provided that they are known to be present in such a manner that employees may be exposed under normal conditions of use or in a foreseeable emergency.

(6) Some employers will voluntarily keep records of training sessions. These could be helpful to CSHOs in assessing compliance with the standard.

(7) Employers are required to ensure that training is provided. Employees may be trained by unions, in trade schools, etc. The employer is responsible for ensuring they have been properly trained. If outside training sessions are used to satisfy this requirement, and the CSHO determines that the employee has not been adequately trained, the employer is subject to citation.

b. **Citation Guidelines**. Citations shall be issued under paragraph (h) of the standard when training is found to be inadequate through program review, discussion with management and employee interviews. The employer is always ultimately responsible for ensuring that employees are adequately trained, regardless of the method relied upon to comply with the training requirements.

9. **Trade Secrets, Paragraph (i)**. Only specific chemical identities may be withheld under the HCS trade secret provisions. Even when a chemical's identity is rightfully withheld as a trade secret, its release may be required by the trade secret access provisions in paragraph (i).

    a. **Inspection Guidelines**. CSHOs evaluating the MSDS and hazard determination programs may request disclosure of trade secret identities under paragraph (i)(12) of the HCS. OSHA shall take all steps feasible to protect trade secret identities, including secure filing and return of information when its use is complete.

    b. **Citation Guidelines**. Where OSHA believes that the chemical manufacturer, importer or employer will not be able to support the trade secret claim, the withholding of a specific chemical identity shall be cited as a violation of paragraph (g)(2). Where OSHA does not question the claim that a specific chemical identity is a trade secret, but the employer has failed to comply with paragraph (i)(1)(i), (ii), (iii) or (iv), or with (i)(2) or (3), such failure shall be grouped with 1910.1200 (g)(2), stating the deficiency in the AVD. For example, the employer claims a trade secret exists but failed to indicate on the MSDS that the specific chemical was being withheld for that reason, as required under paragraph (i)(1)(iii).

10. **Response to Medical Emergencies.** The HCS permits a treating physician or nurse to designate the existence of a medical emergency requiring the immediate disclosure of trade secrets.

a. **Inspection Guidelines.** Referrals received from treating physicians and nurses relating to a medical emergency shall normally be classified as imminent danger or serious in accordance with the FOM, Chapter IX. Due to the potential risk to life and/or health, the Area Director shall ensure that these referrals are processed as soon as received. The Area Director or his/her designee shall contact the manufacturer of the chemical by telephone. Telephone numbers are required on the MSDS. The manufacturer shall be informed of the standard's requirements and requested to immediately provide the needed information directly to the treating physician or nurse.

b. **Citation Guidelines.** Failure to disclose the information shall result in the issuance of a willful citation. The chemical manufacturer will frequently be located under a different Area Office jurisdiction. Apparent violations shall be referred to the office of jurisdiction for investigation and the issuance of citations. Concurrently, the Area Director of jurisdiction shall coordinate obtaining an administrative subpoena ordering the immediate disclosure of the needed information. Federal Court Orders shall be sought immediately if the administrative subpoena is not effective in obtaining the information.

11. **Response to Nonemergency Referrals.** When health professionals providing medical or other occupational health services to exposed employees, or when employees themselves and/or their designated representatives are denied access to trade secret information, the matter may be referred to OSHA for enforcement proceedings.

    a. As stipulated in the standard, OSHA should receive from the referring health professional, employee, or designated representative a copy of the written request for the trade secret information, as well as a copy of the written denial provided by the holder of the trade secret. These two written documents shall be reviewed by the Area Director to determine the validity of the request and the trade secret claim. The Regional Solicitor will provide assistance in this regard.

    b. If the Area Director does not believe that there is enough information upon which to base a decision, he/she may contact either the trade secret requester or the trade secret holder for further information. Such requests shall be documented in the case file.

L. <u>Classification and Grouping of Violations.</u> The procedures in the FOM, Chapters IV, C.8., and V, C., shall be followed except as modified by this instruction; however, if deviations appear appropriate, they shall be coordinated with the Directorate of Compliance Programs, Office of Health Compliance Assistance, through the Regional Office. The following guidelines normally shall apply:

1. Citations for violations of paragraphs (e), (f), (g) and (h) of the standard shall be issued as separate items when there is a complete lack of a hazard communication program. Otherwise the guidance provided in the FOM or specific guidance in this instruction shall be followed.

2. Serious violations shall be issued whenever a deficiency in the program can contribute to a potential exposure capable of causing death or serious physical harm. In addition, the CSHO must document that the employer knew or should have known of the violation.

    a. Serious violations should be considered only when there is documentation which demonstrates that the employer or downstream employer is using the chemical in a manner which could result in actual or potential exposure capable of producing death or serious physical harm. The lack of a label or the lack of a training program alone for a specific chemical or type of hazard could result in a situation where exposure to that hazardous chemical without these safeguards of the HCS would create a serious hazard.

b. Documentation of a HCS violation for a chemical manufacturer or importer could be in the form of a referral generated as a result of OSHA's observation of conditions of use resulting in employee exposure to the hazardous chemical at a downstream user's workplace.

3. Willful violations should be considered in accordance with the guidelines of the FOM; i.e., the employer committed an intentional and knowing violation of the Act.

   a. The employer was aware that a hazardous condition existed and did not make a reasonable effort to eliminate the condition, and

   b. The employer was aware that the condition violated a standard and was aware of the standard.

   c. In addition, willful citations shall be issued when an employer refuses to provide specific chemical identity information in a medical emergency (29 CFR 1910.1200(i)(2)).

M. <u>Interface With Other Standards.</u> In some cases, an *
employer's duties under other OSHA standards dovetail with requirements of the HCS, resulting in simplified compliance.

1. <u>Medical Records Access</u>. The Access to Employee Exposure and Medical Records standard (29 CFR 1910.20) and the HCS overlap with regard to MSDSs. MSDSs are specifically identified as exposure records under 29 CFR 1910.20(c)(5)(iii). Each the MSDS received by an employer must be maintained for at least 30 years as required at 1910.20(d)(1)(ii). The access standard does offer an alternative to keeping the MSDSs at 1910.20(d)(1)(ii)(B), which reads as follows:

   Material safety data sheets and paragraph (c)(5)(iv) records concerning the identity of a substance or agent need not be retained for any specified period as long as some record of the identity (chemical name if known) of the substance or agent, where it was used, and when it was used is retained for at least thirty (30) years.

Therefore, an employer may discard the original data sheet and retain only the new data sheet if a record of the original formulation is maintained.

a. Paragraph (e)(1)(i) of the HCS requires that employers maintain a list of hazardous chemicals as part of the written hazard communication.

b. Employers might simplify their responsibilities as they relate to the overlap between these two standards by incorporating the requirements under 1910.20(d)(1)(ii)(B) with those for the HCS paragraph (e)(1)(i). That is, the list of hazardous chemicals could include information on where chemicals were used and when they were used. These lists would then have to be kept for at least 30 years.

c. Section (e)(4) of the HCS requires employers to make the written hazard communication program available upon request to employees, their representatives, OSHA or NIOSH, in accordance with the requirements at 1910.20 (e). The standard, 1910.20 (e), requires the employer to provide a copy of the requested record (in this case, a copy of the written hazard communication program) "in a reasonable time...but in no event later than fifteen (15) days...." Some employers have incorrectly interpreted this to mean that they have 15 days to produce a copy of the written program and make it available at the worksite. This is an _incorrect_ interpretation; the intent behind the (e)(4) requirements of the HCS is to allow the employer up to 15 days to provide a written (photo or other) copy of the program to employees who request it. This does _not_ mean the employer has 15 days in which to get the program to the worksite for employees to access. The written program must be available to employees at the worksite at all times, as per 1910.1200 (e)(1). (See Appendix A, discussion at (e)(2) page A-15.)

2. _Air Contaminants_. OSHA enforcement of the new Air Contaminants rule was effective September 1, 1989.

Chemical manufacturers, importers, distributors or employers who prepare MSDS were responsible for incorporating the changes precipitated by the new standards within three (3) months. Therefore, all MSDS and labels must already have been modified if affected by this rulemaking and such modifications of both PEL (including STEL and skin notations) and health hazard data must now appear on the revised MSDS and labels, as appropriate.

3. <u>29 CFR 1910.1450, Occupational Exposure to Hazardous Chemicals in Laboratories</u>. Quality control laboratories are usually adjuncts of production operations and are <u>not</u> covered under the Laboratory Standard, but rather would be covered under the HCS. For other laboratories covered under the Laboratory Standard, the requirements of the HCS are superseded (the more specific standard, 1910.1450, takes precedence). Both the training and information and the hazard identification requirements of the Laboratory Standard are more extensive than the HCS laboratory requirements.

4. <u>Other Health Standards</u>. Paragraph (f)(4) of the HCS references labeling requirements of substance-specific standards. Employers must comply with these substance specific standards. For example, the ethylene oxide (ETO) standard provides a different labeling requirement than the HCS. Labels do not have to be affixed to containers of ETO unless the product is capable of producing employee exposure at or above the action level of 0.5 ppm as an 8-hour time weighted average (29 CFR 1910.1047 (j)(1)(ii)).

N. <u>Evaluation</u>. In keeping with agency policy, an evaluation of the effectiveness of this instruction shall be conducted annually. An evaluation report from each Region shall be written and submitted to the Directorate of Compliance Programs within 30 days of the close of the fiscal year. Elements to be considered in the evaluation are the following:

1. Are enforcement and citation policies clear?

2. Are particular problems not addressed or inadequately addressed in this instruction?

3. Are parts of this instruction not useful?

[signature]

Gerard F. Scannell
Assistant Secretary

DISTRIBUTION: National, Regional, and Area Offices
                Compliance Officers
                State Designees
                NIOSH Regional Program Directors
                7(c)(1) Project Managers

Appendix A

<u>Clarifications and Interpretations of the Hazard Communication Standard (HCS)</u>.

This appendix includes clarifications and interpretations which respond to the most frequently asked questions and points of common misunderstanding. Where possible, clarifications are keyed to the most applicable paragraph of the HCS. In many cases a clarification applies to an entire paragraph of the standard. These are included after each section.

<u>Purpose</u>.

(a)(2)   OSHA's position is that State standards can be enforced only under the auspices of an OSHA-approved State plan. States without State plans are preempted from addressing the issue of Hazard Communication. Community right-to-know standards are outside the jurisdiction of OSHA and are not affected by this position. Inquiries regarding preemption that require in depth knowledge of this subject shall be referred through the Directorate of Compliance Programs to the Office of State Programs for response.

The agency's position regarding State standards has been described in OSHA Instruction STP 2-1.117. This should be consulted when answering questions regarding such State standards.

<u>Scope and Application</u>.

(b)(1)   The HCS has a unique requirement for downstream disclosure of information from chemical manufacturers and importers to employers receiving their products. This downstream flow of information is essential to the complete implementation of the standard, but does create enforcement situations that have not been encountered with previous standards. The CSHO's familiarity with the procedures established in this instruction to address such situations is essential to implementation of the HCS.

(b)(2)   The phrase "known to be present" is essential to understanding the scope of the standard. If a hazardous chemical is known to be present by the chemical manufacturer or the employer, it is covered by the standard. This includes chemicals to which employees may be exposed during normal operations or in a foreseeable emergency. This means that even though an employer did not create the hazard, such as silica exposure during concrete demolition, or the hazards of exposure to the chemicals brought onto a multi-employer worksite by other employer(s), the standard applies and the employer whose employees are exposed to chemicals known to be present should include hazard communication information about these exposure situations in his workplace hazard communication program.

By-products are also covered by the HCS. Employers' hazard determination procedures must anticipate the downstream use of their products and account for any hazardous by-products which may be formed. For example, a manufacturer of gasoline must inform downstream users of the hazards of carbon monoxide, since carbon monoxide is a hazardous chemical and is a "known to be present" by-product resulting from the use of gasoline. Similarly, manufacturers of diesel fuel must inform downstream users of the potential human carcinogenicity of diesel exhaust on the MSDSs for diesel fuel. (See NIOSH <u>Current Intelligence Bulletin</u> No. 50, August, 1988.)

The terminology "exposed under normal conditions of use or in a foreseeable emergency" excludes products or chemicals that do not meet this condition. For example, a chemical that is inextricably bound in a mixture and presents no potential for exposure would not be covered. This paragraph must be read in conjunction with the definition of exposure which specifically includes potential (either accidental or possible) exposure. (See the FOM for guidance on citing potential exposure.) Further, employees such as office workers who encounter chemicals only in non-routine, isolated instances are not covered. However, an office worker who works in a graphic arts department and routinely uses paints, adhesives, etc., would be covered by the HCS.

OSHA has never considered either radioactivity or biological hazards to be covered by the HCS. If, however, another type of hazard is presented along with the material (e.g., a container with a biological sample packed in a hazardous solvent), then the container would be subject to the requirements of the HCS for the other hazardous chemical.

(b)(3) The coverage of laboratories is limited under the HCS. Although the standard does not specifically define the term "laboratory", it is intended to mean a workplace where relatively small quantities of hazardous chemicals are used on a nonproduction basis; i.e., bench-scale operations. The definition would include research facilities as well as quality control laboratory operations located within manufacturing facilities. Establishments, however, which produce samples or chemical standards to be sent out to other employers covered by the HCS would **not** fall under the standard's term for a laboratory. Those employers who ship hazardous chemicals would be considered either chemical manufacturers or distributors and must label in accordance with paragraph (f)(1) and provide MSDS per paragraphs (g)(6) and (g)(7).

29 CFR 1910.1450, Exposure to Hazardous Chemicals in Laboratories, addresses hazard communication requirements in laboratories. It is consistent with the HCS, but also has some additional requirements that must be applied in laboratories covered by that rule. The operating definition of a laboratory is not the same for both standards. 29 CFR 1910.1450 covers only laboratories meeting criteria of "laboratory use" and "laboratory scale" and excludes procedures that are part of a production process (55 FR 3328). The preamble to 29 CFR 1910.1450 states "... most quality control laboratories are not expected to meet the qualification for coverage under the Laboratory Standard. Quality control laboratories are usually adjuncts of production operations..." (55 FR 3312). Quality control laboratories would therefore generally be covered by the HCS.

Under the HCS, laboratories do not have to have a written hazard communication program. Therefore, when the required training is performed, the part that deals with the program availability will simply point out that such written programs are not required for laboratories.

Some manufacturers of chemical specialty products have interpreted the laboratory provisions as exempting them from coverage. These operations are considered to be manufacturing processes, and are not exempted. Furthermore, a pilot plant operation is also considered to be a manufacturing operation, not a research laboratory operation. In addition, establishments such as dental, photofinishing, and optical laboratories clearly are not considered laboratory operations for the purposes of this standard since they are engaged in the production of a finished product.

Quality control samples taken in a plant must be labeled, tagged, or marked unless the person taking the sample is also going to be performing the analysis, and thus the sample would come under the portable container exemption. A hand-written label may be utilized as long as required label information is present. The rack in which samples are placed could be labeled in lieu of labeling individual samples if the contents and hazards are similar.

(b)(4) Since all containers are subject to leakage and breakage, employees who work in operations where they handle only sealed containers (such as warehousing) are potentially exposed to hazardous chemicals and therefore need access to information as well as training. The training required for employees who handle sealed containers is dependent upon the type of chemicals involved, the potential size of any spills or leaks, the type of work performed and what actions employees are expected to take when a spill or leak occurs.

Employers are required to obtain a MSDSs for chemicals in sealed containers if an employee requests one. The employer's attempt must begin promptly (within a day) in order to be consistent with the requirement that available sheets be accessible during each shift in the work area.

(b)(5) These exemptions apply to labeling requirements of the HCS only and are not intended to provide a complete exemption from the standard.

(b)(6) This paragraph totally exempts certain categories of substances from coverage under the HCS. Hazardous waste is completely exempted from the standard when subject to regulation by the Environmental Protection Agency (EPA), under the Resource Conservation and Recovery Act (RCRA). If the waste is not regulated under RCRA, then the requirements of the standard apply. Once the material is designated as hazardous waste as defined under RCRA, it is totally exempted. Other chemicals which are used by employees at a hazardous waste site that are not hazardous waste are covered under the HCS. (An example would be an acid brought on site by the employer to neutralize a waste product.)

Under the current rule, whenever a consumer product is used in a manner that is not comparable to typical consumer use, it is covered by the HCS. The standard requires the employer to ascertain whether the workplace use is more frequent, or of longer duration than would be expected in normal consumer use. Exposures in these situations would be greater, and thus the need increases for additional information for employee protection. The use of cans of spray paint during production runs rather than for occasional, short, one-time applications that typify consumer use is an example of hazardous chemical use which would not qualify as consumer product use.

The key to the definition of "article," and thus the exemption, is the term "under normal conditions of use." For example, an item may meet the definition of "article," but produces a hazardous by-product if cut or burned. If the cutting or burning or otherwise processing the article in such a way as to result in employee exposure to a hazardous chemical is not considered part of its normal conditions of use, the item would be an "article" under the standard, and thus be exempted.

As mentioned in the preamble to the August 24, 1987 rule, exposures to releases of "very small quantities"; e.g., a trace amount, are not considered to be covered by the HCS. Thus, absent evidence that releases of such "very small quantities" could cause health effects in employees, the article exception to the rule's requirements would apply. The following items are examples of articles:

> Stainless steel table
> Vinyl upholstery
> Tires
> Adhesive tape

The following items are examples of products which would NOT be considered "articles" under the standard, and would thus not be exempted from the requirements:

> Metal ingots that will be melted under normal conditions of use.
>
> Bricks for use in construction operations, since, under normal condition of use, bricks are cut or sawed, thereby resulting in exposure to crystalline silica.
>
> Switches with mercury in them that are installed in a maintenance process when it is known that a certain percent break under normal conditions of use.
>
> Lead acid batteries which have the potential to leak, spill or break during normal conditions of use, including foreseeable emergencies. In addition, lead acid batteries have the potential to emit hydrogen which may result in a fire or explosion upon ignition.

It should be noted that the only information that has to be reported in these situations is that which concerns the hazard of the released chemical. The hazardous chemicals which are still bound in the article would continue to be exempted under the "article" exemption.

The wood and wood products exemption was never intended by OSHA to exclude wood dust from coverage. This fact was clarified in the preamble to the final rule published August 24, 1987. (See <u>Federal Register</u>, Vol. 52, No. 163, page 31863.) The permissible exposure limits for wood dust recently adopted under OSHA's PEL Project must be included on the MSDSs, which will generally be developed by the sawmill. Further, any chemical additives present in the wood which represent a health hazard must also be included on the MSDSs and/or label as appropriate.

<u>Definitions</u>.

(c) The definitions of the HCS must be heavily relied upon to properly interpret and apply the standard. In many cases terms within a definition are themselves defined within the same section.

<u>Article</u>. The definition has been interpreted to permit the release of very small quantities of a hazardous chemical and still qualify as an article provided that a physical or health risk is not posed to the employees. Examples of very small quantities would be the release of a few molecules or trace amounts of a hazardous chemical (52 <u>FR</u> 31865).

<u>Chemical Manufacturer</u>. Based on this definition and that of its related terms, an employer that manufactures, processes, formulates, or repackages a hazardous chemical is considered a "chemical manufacturer." This definition includes someone who blends or mixes chemicals; such persons may comply with the standard by merely transmitting the relevant label/MSDSs for the ingredients, which they received in good faith from their suppliers, to their downstream customers. Oil and gas producers are chemical manufacturers for the purposes of the HCS because they process hazardous chemicals for use or distribution.

For substances which are grown, cultivated, or harvested and which are not processed by the grower before being sold, the first employer meeting the definition of "chemical manufacturer" will be responsible for performing the hazard determination, developing or obtaining the MSDSs, and labeling containers of the hazardous chemicals. For example, saw mills and grain elevators will be considered to be the "chemical manufacturer" since they are the first employers who meet the definition. A saw mill processes timber into lumber (meets definition of "produce") thereby creating wood dust in the process, which is a hazardous chemical under the HCS. Grain elevators will also meet the definition of a "chemical manufacturer" since they treat, dry, and move grain, creating grain dust (which is also a hazardous chemical under the standard).

Commercial Account. A commercial account is an arrangement whereby a retail distributor sells hazardous chemicals to an employer, generally in large quantities over time and at costs that are below the regular retail price.

Container. This definition includes tank trucks and rail cars. A room or an open area is not to be considered a container and, therefore, a hazardous chemical such as wood dust on the floor of a workplace, or a pile of sand at a construction site, would not have to be labeled. Since only "containers" need to be labeled under the HCS, if there is no container, there is no requirement to label.

Pipes or piping systems, engines, fuel tanks, or other operating systems in a vehicle are not considered to be containers. Thus, LP cylinders that serve as the source of fuel used to operate lift trucks, for example, would not have to be labeled once the fuel tank is installed, although the spare LP cylinder(s) in storage must be labeled since they are containers. Although containers of fuel such as gasoline and LP clearly are within the scope of the HCS, no requirement exists to label the lift truck. The producer still has an obligation to assess the hazards associated with the fuels, including their by-products.

The standard requires all containers of hazardous chemicals leaving the workplace to be labeled with the required information. Even very small containers must be tagged or marked in a fashion that fulfills the intent of the standard.

Distributor. A distributor who blends, mixes or otherwise changes the chemical composition of a chemical is to be considered a chemical manufacturer under the HCS. As a result, employees in those operations are to be considered just like other employees who use hazardous chemicals. A distributor, therefore, performing a chemical manufacturing operation (i.e., blending, mixing, etc.) becomes a chemical manufacturer and will probably need to give additional training to those employees performing the manufacturing operation since the distributor will not be able to satisfy the sealed container provision in paragraph (b)(4) and invoke its limited requirements.

Employee. Employees, such as office workers or bank tellers who encounter hazardous chemicals only in non-routine, isolated instances are not covered. For example, a worker who occasionally changes the toner in a copying machine would not be covered by the standard. However, an employee who operates a copying machine on a full-time basis would be covered by the provisions of the HCS for any hazardous chemicals used.

Exposure. It is important to note, especially for purposes of chemical manufacturers' hazard determinations, that "exposure" includes any route of entry (inhalation, ingestion, skin contact or absorption) and includes potential (accidental or possible) exposure including exposure that could result in the event of a foreseeable emergency.

Hazard Determination.

(d)(1) Although the chemical manufacturer and the importer have the primary duty for hazard evaluation, it is expected that some employers will choose to do their own evaluations. Whoever does the evaluation is responsible for the accuracy of the information. The evaluation must assess the hazards associated with the chemicals including those hazards related to any anticipated or known use which may result in worker exposure.

Known intermediates and by-products are covered by the HCS. Decomposition products which are produced during the normal use of the product or in foreseeable emergencies (e.g., plastics which are injection molded, diesel fuel emissions) are covered if the hazardous chemicals are known to be present. "Foreseeable emergency" does not include employee exposures in the event of an accidental fire, but does include equipment failure, rupture of containers, or failure of control equipment which could result in an uncontrolled release of a hazardous chemical.

An employer may rely upon the hazard determination performed by the chemical manufacturer. Normally, the chemical manufacturer possesses knowledge of hazardous intermediates, by-products and decomposition products that can be emitted from his chemical product. However, if the employer obtains information regarding the hazards from a source other than the manufacturer, the employer is responsible for including such information in his hazard communication program.

(d)(2) The preparer of the MSDSs/labels is required to consider all available scientific evidence concerning the hazard(s) of a chemical in addition to consulting the floor reference sources listed in paragraph (d)(3) of the standard. (See Appendix C of this instruction for further guidance on evaluating health effects.) No testing of chemicals to determine hazards is ever required; the evaluation is to be based on information currently available in the literature.

Where at least one positive scientific study exists which is statistically significant and demonstrates adverse health effects, the MSDSs must include the adverse health effects found. This does not necessarily mean that the results of all such studies would also appear on the label.

The standard's definition of "chemical" is much broader than that which is commonly used. Thus, steel coils which are cut and processed, castings which are subsequently ground or welded upon, carbide blades which are sharpened, and portland cement, which is both a skin and eye irritant, are all examples of chemicals which would normally be covered since exposure to hazardous chemicals would occur in the workplace.

Any substance which is inextricably bound in a product is not covered under the HCS. For example, a hazard determination for a product containing crystalline silica may reveal that it is bound in a rubber elastomer and under normal conditions of use or during foreseeable emergencies cannot become airborne and therefore cannot present an inhalation hazard. In such a situation, the crystalline silica need not be indicated as a hazardous ingredient since it cannot result in employee exposure.

(d)(3) Any compound of a substance regulated in part 1910, Subpart Z, including those listed in the Z Tables or for which there is a TLV in the latest edition of the ACGIH, Threshold Limit Values listing, is considered to be part of the floor of hazardous chemicals covered by the standard.

<u>Nuisance Dust or Particulates</u>. The term "nuisance dust" is no longer used in 1910.1000. A number of particulates now have specific PELs and are covered by the HCS. The particulates not otherwise regulated are exempt unless evidence exists that they present a health or physical hazard other than physical irritant effects. For these chemicals, the "Particulates not otherwise regulated" PELs must be included on the MSDSs.

(d)(4) On December 20, 1985, OSHA published an interpretive notice in the <u>Federal Register</u> regarding the carcinogenicity of lubricating oils (VOL. 50 FR 51852). The notice was published in response to a number of inquiries which were received regarding the applicability of the HCS requirements to naphthenic lubricating oils which are refined using a hydrotreatment process. These types of oils may be found in a number of industrial operations, including ink manufacture and the production of synthetic rubber.

Positive findings of carcinogenicity by the International Agency for Research on Cancer (IARC) must be reported under the HCS. The IARC Monograph 33 concludes that there is sufficient evidence to indicate that mildly hydrotreated and mildly solvent refined oils are carcinogenic. Therefore, under the requirements of the HCS, producers of such materials must report such findings on the MSDSs for the substance and include appropriate hazard warnings on labels.

IARC also stated that there is inadequate evidence to conclude that severely hydrotreated oils are carcinogenic, and that there is no evidence to indicate that severely solvent-refined oils are carcinogenic. In the absence of any valid, positive evidence from sources other than IARC regarding the carcinogenicity of severely hydrotreated or severely solvent-refined oils, no reference to carcinogenicity need be included on the MSDSs and labels for such materials. IARC has also concluded that when an oil is refined using sequential processing of mild hydrotreatment and mild solvent refining, there is no evidence of carcinogenicity.

The questions posed to OSHA concerned the process parameters used for mild hydrotreatment. OSHA examined the studies upon which IARC based its positive findings of carcinogenicity to determine the process parameters used to refine the oils studied. Any oil will be considered to be mildly hydrotreated if the hydrotreatment process was conducted using pressures of 800 pounds per square inch or less, and temperatures of 800 degrees Fahrenheit or less, independent of other process parameters. If the oil is being produced within the specified parameters, it must be considered to be potentially carcinogenic under the requirements of the HCS.

It should also be noted that negative evidence generated by a producer does not negate the positive IARC finding and cannot be used to dispute positive findings relating to any substance. The producer is free to report any negative findings as well, but there is a positive duty to report IARC's conclusions.

(d)(5) While the HCS does not require testing of chemicals to determine their hazards, some preparers of MSDSss are apparently considering testing mixtures as a whole so as not to have to list individual hazardous ingredients on the MSDSs. Should employers choose to pursue this option; i.e., to test the mixture as a whole, a full range of tests would have to be performed, including tests to determine health hazards (acute and chronic) and physical hazards. Employers may also choose to test for certain hazards or properties and rely on the literature for published information on the other hazards. Compliance officers can expect to see MSDSss which use both the tested and untested mixture approaches; e.g., perhaps an employer has determined a flashpoint for the mixture, but has not tested it for health hazards but has relied instead on information in the published literature for this section of the MSDSs. Such an approach to hazard determination is acceptable under the HCS. Where the physical characteristics have not been objectively determined, the employer may present data on the components in ranges; e.g., flash points range from 70 to 100 degrees fahrenheit.

(d)(6) Employers who are not planning to evaluate the hazards of chemicals they purchase can satisfy the requirement for written hazard evaluation procedures by stating in their written program that they intend to rely on the evaluations of the chemical manufacturer or importer.

Downstream employers/employees do not have access to the written procedures maintained by the chemical manufacturer/importer. If there appears to be a problem with the information received, and it cannot be resolved with the supplier of the product, the matter should be referred to OSHA for investigation. OSHA does have access to the written procedures.

Written Hazard Communication Program.

(e)(1) All employers with employees who are, or may be, exposed to hazardous chemicals known to be present in their workplaces, must develop, implement, and maintain at primary workplace facilities and fixed worksite locations a written hazard communication program. Programs must be developed whether the employer generates the hazard or the hazard is generated by other employers. An effective program is one that promotes the safe handling and use of hazardous chemicals in the workplace.

(e)(2) Although a multi-employer worksite is not defined in the HCS, it is intended to mean those establishments where employees of more than one employer are performing work and are exposed to hazardous chemicals. The MSDSs information exchange or access requirements pertain to employers who introduce hazardous chemicals into the worksite and expose another employer's employees.

All types of worksites may be "multi-employer worksites," not just construction sites. For example, a manufacturing employer becomes the "exposing employer" if he produces, uses or stores chemicals in such a way that he may expose the employees of another employer to hazardous substances. Now that the HCS is in effect in all industry sectors, an exposing employer must advise outside contractors working at his plant about the hazardous chemicals that the contractor's employees may be exposed to and vice versa.

Paragraph (e)(2)(i) requires an employer on a multi-employer worksite to provide other employers with a copy of pertinent MSDSs or to make them available at a central location in the workplace. This requirement covers each hazardous chemical to which the other employer's employees may be exposed. Therefore, one employer does not actually have to physically give another employer the MSDSs, but the employer must inform the other employer of the location where the MSDSs will be maintained (e.g., in the general contractor's trailer). The performance-orientation of the rule allows employers to decide the method to be used to accomplish the required exchange of information.

In the construction industry, it would probably be most efficient for the general contractor to coordinate the requirement for maintaining MSDSs on site. For example, the general contractor could keep and make available MSDSs in the office on the site.

An employer must provide MSDS(s) to other employers or make them available in a central location if the other employers will have employees exposed or potentially exposed. The potential exposure could even occur at some time in the future. For example, if a painting contractor's workers are using flammable solvents in an area where another subcontractor's workers are welding pipes, then the painting contractor must ensure that the MSDSs for the flammable solvents are available to the welding subcontractor's employees. However, if electricians are not working near or at the same time as the painting contractor, and therefore it is not possible for either employer's employees to be exposed, then no exchange of MSDSs is required.

The HCS's "multi-employer workplaces" provision at (e)(2) states that employers who produce, use or store hazardous chemicals at a worksite in such a way that the employees of other employers may be exposed must include in their written hazard communication program the methods to ensure that the other employers are adequately informed of the hazards and appropriate precautionary measures to be taken so they can protect their own employees.

The intent of the HCS is met on multi-employer worksites when information on the hazards of chemical substances at the worksite is available to all affected employers and employees. All employers with employees potentially exposed to hazardous chemicals therefore must have in place an effective written hazard communication program that details how this intent will be met.

If an employer does not bring hazardous chemicals on site, a list of hazardous chemicals is not required as part of his hazard communication program. Nevertheless, the employees must be trained how to use labels and MSDSs, to recognize hazards and to follow appropriate protective measures.

An exception to the requirement that the written hazard communication (HCP) be kept on-site on multi-employer worksites may be found in situations where an employee or employees must travel between workplaces or where their work is carried out at more than one geographical location, yet who, at some time, report to a primary workplace facility where the written HCP is maintained. The standard sets forth, at (e)(1), a positive requirement for the written program to be maintained "at the workplace." OSHA has interpreted this requirement to mean that the written program must be kept on-site, at all times, or even in the work truck of employees who travel between worksites.

However, the Agency has proposed, in the 1988 Notice of Proposed Rulemaking, to add a new subparagraph to the paragraph (e) requirements to allow the written program to be maintained at a "central location at the primary workplace facility" for employees who travel between workplaces during a workshift (proposed new paragraph (e)(5)). The final rule presently allows MSDSs to be maintained at the central workplace for employees who travel between workplaces during a workshift (paragraph (g)(9)). The (g)(9) provisions also require that employees have immediate access to information in an emergency which is important since MSDSs must be readily accessible to employees in the event of an emergency, accidental leak, spill, etc.

Unlike the immediate need for MSDSs information to be readily accessible to employees while they are in their work area(s), the information contained in an employer's written HCP is mainly procedural and its presence on the worksite other than a fixed location may not have a direct or immediate relationship to employee safety or health. This is especially true in situations where employers are implementing an effective overall HCP and whose employees have already received the required hazard communication training. This means that employees are aware of the requirements of the employer's HCP, including being familiar with the list of hazardous chemicals known to be present, the labeling system in use, the presence of and accessibility to MSDSs, and have been trained in accordance with paragraph (h) requirements. The need for the program to be on-site, therefore, in situations where employees travel or are dispatched from a primary workplace location (e.g., administrative offices) where the written program is maintained to a multi-employer worksite may bear no immediate relationship to safety and health and may, in the professional judgment of the CSHO and Area Director, be considered a "de minimis" violation of section (e)(1). (See the FOM, Chapter IV, B.6., pages IV 30-31.)

This citation policy change applies even in situations where the employee does not return to the primary workplace during the workshift as long as the employee(s) is aware of the content of the program and the methods the program contains that affect the sharing of the hazard communication information required at (e)(2)(i-iii). Stated in another way, if hazard communication information (accessibility of MSDSs, the employer's labeling system, etc.) is not being shared with other on-site employers and the employees are unaware of the methods outlined in the program which have been developed to accomplish this intent, then the need for the program to be on-site would bear a direct relationship to safety and health and the absence of the program on-site would not be a "de minimis" violation.

At fixed worksite locations, the requirement for the written hazard communication program to be maintained on-site and readily accessible to employees remains. Again, an effective program is one that promotes the safe handling and use of hazardous chemicals in the workplace. Its immediate presence in other than fixed worksite locations bears a direct relationship to safety and health only when its procedural direction is necessary to direct the employers in their implementation of the overall hazard communication program's requirements.

OSHA's compliance and enforcement policies for multi-employer worksites are set forth in the FOM, Chapter V, sections F.1. and 2., which state that, with regard to working conditions where employees of more than one employer are exposed to a hazard, the employers "with the responsibility for creating and/or correcting the hazard" shall be cited for violations of OSHA standards that occur on that multi-employer worksite. Normally citations for violations shall be issued to each of the exposing employers as well as to the employer responsible for correcting or ensuring the correction of the condition.

Whenever the general contractor or the construction manager on a multi-employer worksite is in the best position to ensure that all contractors on site with hazardous materials comply with the standard's requirements, the general contractor or construction manager shall be cited for violations of the HCS as well as any contractor who has not complied.

(e)(4) Paragraph (e)(4) requires employers to make the written program available upon request to employees, OSHA and NIOSH, in accordance with the requirements of the Access Standard, 29 CFR 1910.20(e). This requirement is interpreted to apply to the requirement of the employer to provide a <u>copy</u> of the written program within the time periods discussed in 1910.20 (i.e., no later than 15 days after the request for access is made). It is <u>not</u> meant to allow an employer of a primary workplace facility or a fixed location worksite a 15-day time period in which to make the program available for inspection on-site. For fixed worksites and primary workplace facilities, the written hazard communication program must be maintained on-site at all times. OSHA interprets the 15-day period referenced in (e)(4) to pertain to the length of time the employer has in which to provide a <u>copy</u> of the program to the requesting party. (<u>See</u> discussion at subparagraph (e)(2) of this Appendix for OSHA citation policy regarding the maintenance of written programs on multi-employer or mobile worksite locations.)

<u>Labels and Other Forms of Warning</u>.

(f)(1) and (f)(5) The purpose for labels under the standard is clear. Labels provide an immediate warning to employees of the hazards they may be exposed to and, through the chemical identity, labels provide a link to more detailed information available through MSDSs and other sources. Labels must contain the identity of the chemical, an appropriate hazard warning, and the name and address of the responsible party.

OSHA recognizes that the degree of detail on a label needed to convey a hazard may be different within a workplace where other information is readily available, compared to labels required on shipped containers, where the label may be the only information available.

The standard's preamble recognizes the existence of numerous labeling systems that are currently in use in industry. Examples include the HMIS (Hazardous Materials Information System), NFPA (National Fire Protection Association) and ANSI (American National Standards Institute) systems. Some of these systems rely on a numerical and/or alphabetic codes to convey the hazards. Although these labeling systems may not convey the target organ effects, the intent of the standard is to permit the use of these systems for _inplant_ labeling as long as the written Hazard Communication Program adequately addresses the issue.

Paragraph (e)(1) of the HCS requires employers to include in their written hazard communication program a description of how the training requirements of paragraph (h) will be met, and subparagraph (e)(2)(ii) requires employees to be trained on the physical and health hazards of the chemicals they work with. OSHA has interpreted this to include being apprised of the target organ effects of the hazardous chemicals employees are or may be exposed to while working. The training program must therefore explicitly instruct employees on how to use and understand the plant's alternative labeling systems to ensure that employees are aware of the effects (_including_ target organ effects) of the hazardous chemicals to which they are potentially exposed.

CSHOs must carefully review the overall hazard communication program to ensure its effectiveness in meeting all the requirements of the HCS. One way for CSHOs to determine the effectiveness of the training program, including employee understanding of target organ effects, especially when numerical or other systems are used for in-plant labeling, is through employee interviews. An employer relying on one of the above-mentioned labeling systems may therefore have to augment his hazard communication training program to specifically address the target organ effects that may not be easily discernable from a numerical warning system.

However, for _shipped containers_ the hazard warning must be included on the label and must specifically convey the hazards of the chemical. OSHA has consistently maintained that this includes target organ effects. Casarett and Doull's _Toxicology, the Basic Science of Poisons_ discusses target organs:

Most chemicals that produce systemic toxicity do not cause a similar degree of toxicity in all organs but usually produce the major toxicity to one or two organs. These are referred to as target organs of toxicity for that chemical.

Appendix A of the HCS clearly states that employees exposed to health hazards must be apprised of both changes in body functions and the signs and symptoms that may occur to signal the changes. A label incorporating a rating system is not permitted for shipped containers unless additional label information is affixed to the container. The specific hazards indicated in the standard's definitions for "physical" and "health" hazards are applicable. Phrases such as "caution", "danger", or "harmful if inhaled", are precautionary statements, not hazard warnings. The definition of "hazard warning" states that the warning must convey the hazards of the chemical and is intended to include the target organ effects. If, when inhaled, the chemical causes lung damage, then that is the appropriate warning. Lung damage is the hazard, not inhalation.

There are some situations where the specific target organ effect is not known. Where this is the case, the more general warning statement would be permitted. For example, if the only information available is an LC50 test result, "harmful if inhaled" may be appropriate.

It will not necessarily be appropriate to warn on the label about every hazard listed in the MSDSs. The data sheet is to address essentially everything that is known about the chemical. The selection of hazards to be highlighted on the label will involve some assessment of the weight of the evidence regarding each hazard reported on the data sheet. Assessing the weight of the evidence prior to including a hazard on a label will also necessarily mean consideration of exposures to the chemical that will occur to workers under normal conditions of use, or in foreseeable emergencies. However, this does not mean that only acute hazards are to be covered on the label, or that well substantiated hazards can be left off the label because they appear on the data sheet.

An example of a situation where it may not be necessary to include the presence of a hazardous ingredient in a formulation when developing the product's label follows: Recently, IARC published monograph no. 44, entitled, "Alcohol Drinking." IARC's determination on the carcinogenicity of ethanol is based on chronic exposure to ethanol through human consumption via the <u>drinking</u> of alcoholic beverages, over time. In performing the hazard determination on a product mixture which contains ethanol as one of the hazardous ingredients, a chemical manufacturer must, under the HCS, include mention of ethanol as a hazardous ingredient on the MSDSs, along with the findings as published in the IARC monograph. As part of the hazard determination, manufacturers must consider exposures to the chemical product that would occur under normal conditions of use or in foreseeable emergencies, and toxicity associated with all routes of entry. If a chemical manufacturer were to formulate a product which contained ethanol as part of the mixture, but the product's intended use did not involve exposure through ingestion of the ethanol, the manufacturer could document the intended use and resultant exposure scenarios on the MSDSs but <u>not</u> label the product as a "carcinogen." Again, the information about the carcinogenicity of ethanol would need to appear on the MSDSs, but since exposure under normal conditions of use, etc., would not involve ingestion and since the only evidence calling ethanol a human carcinogen comes from studies involving chronic alcoholic beverage ingestion, the weight of the evidence would preclude the requirement to warn of carcinogenic hazards on the label of the product.

Exposure calculations are not permitted in determining whether a hazard must appear on a label. If there is a potential for exposure other than in minute, trace or very small quantities, the hazard must be included when substantiated as required by the HCS. Suppliers may not exclude hazards based on presumed levels of exposure downstream (i.e., omitting a carcinogenic hazard warning because, in the supplier's estimate, presumed exposures will not be high enough to cause the effect). The hazard is an intrinsic property of the chemical. Exposure determines degree of risk and should be addressed in training programs by the downstream employer.

The labeling requirements for shipped containers leaving the workplace apply regardless of whether the intended destination is interstate or intrastate. If the shipment is to another establishment, even within the same company, the shipped labeling provisions apply. Even sealed containers intended for export must comply with the labeling provisions if these containers leave the workplace and if downstream employees such as dock workers may be exposed to the hazardous chemical(s).

Containers must be labeled as soon as practicable before leaving the workplace. If the container is a tank truck, rail car, or other vehicle carrying a hazardous chemical(s) not already in a labeled container(s), the appropriate label or label information may either be posted on the tank or vehicle, or attached to the accompanying shipping papers or bill of lading. Employers purchasing hazardous chemicals must ensure that their employees are aware of the label warning before potential exposure to incoming chemicals occur. A label may not be shipped separately, even if it is prior to shipment of the hazardous chemical since to do so defeats the intended purpose which is to provide an immediate hazard warning. Mailing labels directly to purchasers will bypass those employees involved in transporting the hazardous chemical. (Note the exception in (f)(2) for solid metals. Containers of solid metals not otherwise meeting the definition of an article need to be labeled only with the initial shipment (unless the information on the label changes).)

Although no explicit requirement exists regarding the updating of labels when new information becomes available, the warning would no longer be appropriate if the MSDSs contained new hazard information that needed to be included on the label. Since the MSDSs must be updated within three months of receipt of new information, the label must be, too, in order to accurately reflect the MSDSs information. Note that distributors have no affirmative obligation to create the container labeling information for hazardous chemicals which they merely send unchanged to their customers, but they do have the responsibility to obtain missing labels from the chemical manufacturer/importer. Distributors must duplicate label information on chemicals which they repackage.

(f)(5) and (f)(6)
An employer's obligation to label in-plant containers of hazardous chemicals requires that all appropriate hazard warnings appear on the label pursuant to (f)(5)(ii). For example, an employer who elects to label only some of the health hazard warnings associated with the chemical while omitting other recognized hazards, such as carcinogenicity, selectively deprives his employees of critical hazard information and shall be cited under (f)(5)(ii). However, if the downstream employer has relied in good faith on the adequacy of the label as prepared by the chemical manufacturer and the label contains an inadequate hazard warning(s), the CSHO shall follow the referral procedures outlined in K.7.a.(7) of this instruction.

For purposes of reviewing alternative in-plant labeling methods under (f)(6), the CSHO shall note that this provision allows alternative means of identification only in the event that an employer chooses to forego labeling an in-plant container under (f)(5). Thus, an employer may not claim that it supplemented its partial compliance with (f)(5)(ii); i.e., labeling only some of the chemical's health hazard warnings, with one of the alternative means of identification enumerated in (f)(6). The key to evaluating the effectiveness of any alternative labeling method is to determine whether it provides an immediate visual warning of the chemical hazards of the workplace, identifies the applicable chemical and container, and conveys the appropriate hazard warnings. The alternative labeling system must also be readily accessible to all employees in their work area throughout each workshift. For purposes of this provision, the term "other such written materials" does not include material safety data sheets used in lieu of labels.

Carcinogen Labeling.

As specified in the rule, chemicals which have been indicated as positive or suspect carcinogens by either OSHA, the International Agency for Research on Cancer (IARC) or the National Toxicology Program (NTP) will be considered to be carcinogenic for purposes of the HCS.

Those chemicals identified as being "known to be carcinogenic" and those substances that may "reasonably be anticipated to be carcinogenic" by NTP must have carcinogen warnings on the label <u>and</u> information on the MSDSs. For NTP, appearing on the annual listing constitutes a positive finding of suspect or confirmed carcinogenicity.

OSHA's comprehensive substance specific regulations in Subpart Z of 1910 contain provisions for labeling. Therefore, containers of hazardous chemicals labeled in accordance with the substance specific standard will be deemed to be in compliance with the health effects labeling requirements of the standard. An exception to this is OSHA's Formaldehyde Standard, for which an administrative stay of the hazard communication provisions (sections (m)(1)(i) and (m)(1)(ii)) is in effect. The HCS is enforceable for these provisions of the formaldehyde standard.

It should be noted that in many instances the labeling requirements of the comprehensive substance specific standard address only carcinogenicity and do not address acute health hazards or physical hazards. Those chemicals regulated by OSHA as carcinogens in substance specific standards that include labeling requirements are listed below:

Asbestos
4-Nitrobyphenyl
Alpha-Napthylamine
Methyl Chloromethyl Ether
3,3' Dichlorobenzidine (and its salts)
Bis-Chloromethyl Ether
Beta-Naphthylamine
Benzidine
4-Aminodiphenyl
Ethyleneimine
Beta-Propiolactone

2-Acetylaminofluorene
4-Dimethylaminoazobenzene
N-Nitrosodimethylamine
Vinyl Chloride (and Polyvinyl Chloride)
Inorganic Arsenic
1,2 Dibromo-3-Chloropropane
Acrylonitrile
Ethylene Oxide
Formaldehyde
Benzene

In addition to those chemicals for which OSHA has substance-specific standards, OSHA has set new permissible exposure limits for several substances based on avoidance of cancer. These substances are specified in the preamble to the Air Contaminants rule published January 19, 1989. (See Table C15-1 on pages 2669-71 of the Federal Register notice of that date.)

IARC evaluates chemicals, manufacturing processes, and occupational exposures as to their carcinogenic potential. The IARC criteria for judging the adequacy of available data and for evaluating carcinogenic risk to humans were established in 1971 (Volumes 1-16) and revised in 1977 (Volumes 17 and following).

The individual monographs contain evaluations on specific chemicals or processes. At the conclusion of each evaluation, IARC provides a summary evaluation for the individual chemical. Periodically, IARC publishes Supplements in which chemicals that have already been evaluated in previous monographs are reevaluated. In cases where a chemical has been reevaluated, the most recent IARC evaluation shall be relied upon.

IARC provides a summary in Supplement 7 of the chemicals which have been evaluated in Volumes 1-42. Table I of Supplement 7 provides a summary evaluation of all chemicals for which human and animal data were considered. Table I of Supplement 7 also provides a summary classification of a chemical's carcinogenic risk:

Group 1 - The agent is carcinogenic to humans.

Group 2A - The agent is probably carcinogenic to humans.

Group 2B - The agent is possibly carcinogenic to humans.

Group 3 - The agent is not classifiable as to its carcinogenicity to humans.

Group 4 - The agent is probably not carcinogenic to humans.

All IARC listed chemicals in Groups 1 and 2A must include appropriate entries on both the MSDSs *and* on the label. Group 2B chemicals need be noted only on the MSDSs.

Individual monographs have been published subsequent to Supplement 7. For purposes of compliance with the MSDSs and labeling requirements, the IARC monograph's summary evaluation for the chemical can generally be relied upon but it may be necessary to review the actual evaluations. In some cases, a group of compounds may be listed in the summary as carcinogenic but closer examination of the appropriate monograph will reveal that IARC had data to support the carcinogenicity of only certain compounds. Those compounds are the only ones covered by the HCS. IARC also evaluates specific industrial processes or occupations for evidence of increased carcinogenicity. Findings that an occupation is at increased risk of carcinogenicity, without identification of specific causative agents, do not affect label or MSDSs requirements.

In addition, the existence of one valid, positive study indicating carcinogenic potential in either animals or humans is sufficient basis for a notation on the MSDSs. Further, if such studies include positive human evidence, then the label must contain carcinogen hazard warnings.

Table 1, below, represents a general guide regarding the labeling and MSDSs requirements under the HCS. The existence of positive human evidence on carcinogenicity always requires carcinogen warnings on the label. In addition, there may be instances where a carcinogen warning may be required for a chemical that is not listed by IARC or NTP but multiple animal studies indicate carcinogenicity. Such cases shall be reviewed by the Regional Administrator and coordinated by the Directors of Compliance and Health Standards Programs.

TABLE 1

GUIDANCE FOR MSDS AND LABEL NOTATIONS FOR CARCINOGENS

Source	MSDS	Label
Regulated by OSHA as a carcinogen	X	X
Listed on NTP Carcinogen Report	X	X
IARC--Group 1	X	X
IARC--Group 2A	X	X
IARC--Group 2B	X	Not Required
IARC--Group 3	Not Required	Not Required
IARC--Group 4	Not Required	Not Required
One Positive Study-Animal Only	X	Not Required
Multiple Animal Studies	X	Depends on weight of evidence; N.O. review needed.
One Positive Study-Some Human Evidence	X	X

Given the above criteria, benzene, which is regulated by OSHA as a carcinogen and for which several valid, positive human studies exist, would require both MSDSs and label notations whereas a substance for which only some animal data exist does not. Polyvinyl resin must be labeled as a carcinogen but final molded and extruded products do not need to be (as per 29 CFR 1910.1017). (See also the discussion on IARC's determination on the carcinogenicity of "Alcohol Drinking," IARC Monograph No. 44, as it pertains to labeling requirements (page A-15).)

Material Safety Data Sheets.

(g)(1) Chemical manufacturers/importers who choose to purchase data sheets for their products from information services, rather than developing them themselves, retain responsibility for providing the sheets and for assuring their accuracy. Employers who in good faith choose to rely upon the sheets provided to them by the chemical manufacturer/importer assume no responsibility for their contents.

The MSDSs requirements apply to free samples provided by chemical manufacturers and importers since the hazards remain the same regardless of the cost to the employer.

Even though solid metals are covered differently under the labeling requirements, the full MSDSs requirements still pertain.

Chemical manufacturers often receive requests for MSDSs from customers for chemicals or articles which are not covered under the HCS. The HCS does not require MSDSs to be provided under those circumstances. If the chemical manufacturer/importer chooses to provide the MSDSs as a customer service, it may be noted on the sheet that the chemical or article has been found by the company not to be covered by the rule. For example:

> This product is not considered to be or to contain hazardous chemicals based on evaluations made by our company under the OSHA Hazard Communication Standard, 29 CFR 1910.1200.

The MSDSs may not indicate that OSHA has made such a finding for the product since the Agency does not make such case-by-case hazard determinations.

The safety and health precautions on the MSDSs must be consistent with the hazards of the chemicals. Some MSDSs include recommendations for protective measures that are for "worst case scenarios," e.g., recommending supplied air suits for products of relatively low toxicity. The HCS requires that accurate information be provided on the MSDSs. This applies as much to "overwarning" on the MSDSs/label as well as the absence of information ("underwarning").

Scrap dealers are generally considered distributors and, since their products are not articles, would NOT be exempt from the HCS. If their suppliers are furnishing articles which they did not manufacture, (such as a broken refrigerator), the supplier is not required to provide a label or MSDSs. However, if their suppliers added hazardous chemicals to the article, as would be the case if an employer scraps pipes that contained a hazardous chemical and continues to contain its residue, the supplier must provide a label and MSDSs to the scrap dealer. In addition, "article" manufacturers that sell for scrap those produced items that fail specification or suppliers who provide, for example, metal tailings from a manufacturing process, are considered by OSHA to have the required knowledge of the item's constituents and must develop and transmit MSDSs and labels to downstream scrap dealers.

(g)(2) The OSHA Form 20 has been obsolete since May 1986. Simply following the titles of the blocks to complete the Form 20 will not result in an appropriate sheet, but it could be modified to comply. Any format is acceptable, as long as the required information is included. OSHA has published a sample MSDSs, form number OSHA-174. This is an optional form which may be used to comply with the HCS.

The requirement that the MSDSs be in English is intended to prevent importers of chemicals from transmitting MSDSs written in a foreign language. However, this requirement was not intended to prevent the translation into foreign languages to aid employee understanding.

If a hazardous chemical is present in the mixture in reportable quantities (i.e., 0.1 percent for carcinogens, and 1 percent for other health hazards), it must be reported unless the mixture has been tested as a whole or unless the material is bound in such a way that employees cannot be exposed. If there really is <u>no</u> exposure (and the standard defines exposure as including <u>potential</u> as well as measurable exposure by any route of entry), either under normal conditions of use or in a foreseeable emergency, then the chemical is not covered by the standard. (See paragraph (b)(2).) In the case of mixtures that are liquid, this provision has to be considered very carefully. For example, if silica is present in a wet mixture it is possible that, if the mixture dries upon application, there is a potential for the silica to become airborne, and thus a potential for exposure. The presence of silica must be indicated on the MSDSs for the liquid mixture in this situation.

For mixtures, if the employer is assuming the mixture has the same hazards as its hazardous components (i.e., no test data on the mixture as a whole), the data sheets for the components will satisfy the requirements of the standard for a data sheet for the mixture. These MSDSs must be physically attached to one another and identified in a manner where they can be cross-referenced with the label. This approach is acceptable provided the MSDSs includes the PEL, TLV, and other exposure limits for <u>each</u> ingredient that has been determined to be a health hazard.

Information must also be included on the MSDSs for ingredients of a mixture present in concentrations of less than 1% (or 0.1% for carcinogens) when the hazardous substance may be released in a concentration which exceeds a PEL or TLV <u>or</u> may present a health risk to exposed employees. An example of the latter may be TDI because it is a sensitizer in very small concentrations, thereby presenting a health risk that must be noted on the MSDSs.

A statement that the chemical is not a carcinogen is not required nor must the MSDSs format include a space for such a statement. However, if the format used provides a space for a carcinogen entry, one must be made since no blank spaces may be present on the MSDSs.

The MSDSs must include a telephone number for emergency information. There is no requirement that the responsible party staff a telephone line with personnel who can respond to an emergency 24 hours a day. The hours of emergency line operation are determined by the chemical manufacturer and should be set after considering the thoroughness of the MSDSs, the hazards of the chemicals, the frequency of use and immediacy of information needs, and the availability of information through alternative sources. One effective alternative used by some suppliers is to have a telephone answering machine that is on when the facility is closed. The message refers callers to the appropriate official in the event of an emergency.

(g)(3) The standard requires that all blocks on a form be completed. Because the standard is performance-oriented, however, employers are free to develop MSDSs in any format they wish (as long as it contains the required information). Computer-generated MSDSs do not have to include fields which do not apply to the chemicals for which it is being used.

(g)(4) Where the evidence can support the fact that a class or family of chemicals presents similar health hazards, it would be appropriate to report those findings on the MSDSs with respect to the entire class or family. Thus, a "generic" MSDSs may address a group of complex mixtures, such as crude oil or natural gas, which have similar hazards and characteristics because their chemical ingredients are essentially the same even though the specific composition varies in each mixture.

(g)(5) Paragraph (g)(5) requires new or significant information to be added to the MSDSs within three months. The Air Contaminants Rule, 29 CFR 1910.1000, was promulgated January 19, 1989, and set new PELs for 164 substances not previously regulated by OSHA and lowered the PELs for 212 substances. These new PELs must appear on the MSDSs. The "old" PELs, referred to as the "transitional limits," air contaminant limits which must be met via the use of engineering controls, may also appear on the MSDSs, but as "new or significant" information regarding the hazard of a chemical, the new PELs must now be included on the MSDSs.

Citations for incomplete or inaccurate MSDSs/labels shall include an abatement requirement for the transmittal of corrected MSDSs/labels to all customers with the next shipment of the chemical.

(g)(6) This paragraph contains the obligation for an employer to obtain the MSDSs as soon as possible if it was not provided with the shipment. It is not necessary for the employer to perform a hazard determination but only to request the MSDSs. If the container label indicates a hazard, the employer will know an MSDSs is necessary.

(g)(6) and (g)(7) Chemical manufacturers and importers have an affirmative duty to provide MSDSs to distributors and employers. Thus, a chemical manufacturer and/or importer shall be cited under (g)(6) if they withhold sending MSDSs to downstream users with an initial shipment or with the first shipment after updating an MSDSs, pending a separate payment for the MSDSs. Similarly, under (g)(7), distributors have an affirmative duty to provide MSDSs to other distributors and downstream employers and cannot withhold sending the MSDSs pending separate payment.

(g)(7) See Definitions (c), in this Appendix, for a discussion of commercial account. Employers purchasing hazardous chemicals from a retail distributor, whose employees will be required to use those chemicals with a greater frequency and duration of exposure than that of regular consumers, must request the MSDS(s) from the retail distributor in order to provide his employees protection under the HCS.

(g)(8) This provision requires MSDSs or electronically accessible MSDSs to be maintained on site. Readable copy of MSDS(s) must be available on-site. This may be accomplished by the use of computers with printers, microfiche machines, and/or telefax machines, any of which would meet the intent of the standard. The key to compliance with this provision is that employees have no barriers to access to the information and that the MSDSs be available during the workshift. When direct and immediate access to paper or hard-copy MSDSs does not exist, CSHOs should evaluate the performance of the employer's system by requesting a specific MSDSs. Mere provision of the requested information orally via telephone is not acceptable.

CSHOs must exercise judgment in enforcing this provision. Factors that may be appropriate to consider when determining if MSDSs are readily accessible may include: Must employees ask a supervisor or other management representative for the MSDSs? Are the sheets or alternative methods maintained at a location and under conditions where employees can refer to them during each workshift, when they are in their work areas? If a computer or FAX system is used, do employees know how to operate and obtain information from the system? Employees must have access to the MSDSs and be able to get the information when they need it, in order for an employer to be in compliance with the rule.

On multi-employer jobsites, employers who produce, use or store hazardous chemicals in such a way that other employers' employees are exposed must also provide copies of or access to MSDSs as discussed in section (e) of this Appendix. Again, actual paper copies of data sheets, computer terminal access, FAX, or other means of providing readable copy on-site are permitted, as long as no barriers to employee access exist.

(g)(9) If employees work at more than one site _during the shift,_ they must be able to immediately obtain the MSDSs information in an emergency. While the MSDSs may be maintained at a central location in the primary workplace facility, a representative of the employer must be available at that central location to respond to requests for emergency information via telephone or other means.

(g)(10) Computerized data sheets are permitted as long as they are readily accessible to employees (i.e., employees have been trained and know how to operate the computers or otherwise access the MSDSs files). Many larger firms use terminals in plant and train key employees to access them. This is acceptable, as long as the information can be obtained during any work shift, as required by the HCS. Similarly, the use of telefax machines to obtain MSDSs is acceptable as long as the system is reliable and readily accessible while employees are in their work areas during all work shifts.

Employee Information and Training.

(h) Employees are to be trained at the time they are assigned to work with a hazardous chemical. The intent of this provision is to have information prior to exposure to prevent the occurrence of adverse health effects. This purpose cannot be met if training is delayed until a later date.

Additional training is to be done whenever a new _hazard_ is introduced into the work area, not a new _chemical_. For example, if a new solvent is brought into the workplace, and it has hazards similar to existing chemicals for which training has already been conducted, then no new training is required. Of course, the substance-specific data sheet must be available, and the product must be properly labeled. If the newly introduced solvent is a suspect carcinogen, and there has never been a carcinogenic hazard in the workplace before, then new training for carcinogen hazards must be conducted in the work areas where employees will be exposed to it.

Complete retraining of an employee does not automatically have to be conducted when an employer hires a new employee, if the employee has received prior training by a past employer, an employee union, or any other entity. It is highly unlikely that no additional training will be needed since employees will need to know the specifics of their new employers' programs such as where the MSDSs are located and details of the employer's in-plant labeling system, if appropriate.

If it is determined that an employee has not received training or is not adequately trained, the current employer will be held responsible regardless of who provided the training to the employee. An employer, therefore, has a responsibility to evaluate an employee's level of knowledge with regard to the training and information requirements of the standard, and the employer's own hazard communication program, including previous training the employee may have received. The training requirements also apply if the employer becomes aware via the multi-employer worksite provision of exposure of his employees to hazards for which they have not been previously trained.

Training need not be conducted on each specific chemical found in the workplace, but may be conducted by categories of hazard (e.g., carcinogens, sensitizers, acutely toxic agents) that are or may be encountered by an employee during the course of his duties. This approach to training may be especially useful when training employees about the types of hazards they may encounter at another employer's worksite.

A frequently overlooked portion of the training provisions is that dealing with emergency procedures. If the chemical is very hazardous, more information would be expected to be provided on the MSDSs and, therefore, the training for emergency procedures, including information about the characteristics of the chemical and precautions to be taken would need to be more extensive. Section 1910.1200(h) requires training of employees on (among other things) the measures employees can take to protect themselves from hazards including emergency procedures and an explanation of the information on the MSDSs. Section (g)(2)(viii) of the HCS requires the MSDSs to address safe handling and use of chemicals which includes cleanup of spills and leaks. Section (g)(2)(x) requires the MSDSs to address emergency and first aid procedures.

Questions have arisen regarding the interface of 1910.120 training requirements for emergency procedures and those for the HCS. The scope and extent of training regarding emergency procedures will necessarily be dependent upon the desired response of employees to an emergency. If the employer intends to merely evacuate the work area, the training in emergency procedures would be quite simple and limited but should include information on the emergency alarm system in use at the worksite and evacuation routes and areas where applicable. However, if the employees are expected to take appropriate action to moderate or control the impact of the emergency in a similar fashion as emergency responders would, then additional training will be required. At a minimum, training these responders on the "emergency procedures" required under section (h) should include, as applicable, leak and spill cleanup procedures, appropriate PPE, decontamination procedures, shut-down procedures, recognizing and reporting unusual circumstances (incidents), and where to go (evacuate to) in an emergency.

Giving an employee a data sheet to read does not satisfy the intent of the standard with regard to training. The training is to be a forum for explaining to employees not only the hazards of the chemicals in their work area, but also how to use the information generated in the hazard communication program. This can be accomplished in many ways (audiovisuals, classroom instruction, interactive video), and should include an opportunity for employees to ask questions to ensure that they understand the information presented to them.

Furthermore, the training must be comprehensible. If the employees must receive job instructions in a language other than English, then training and information will probably also need to be conducted in a foreign language.

Trade Secrets.

(i)(2)　　The designation of an incident as a "medical emergency" is left to the discretion of the treating physician or nurse.

Appendix B

Sample Letter, MSDS/Label Query

Dear (Name or Position of Responsible Employer Representative):

Representatives of the Occupational Safety and Health Administration (OSHA)/or State plan designated agency recently visited/or corresponded with (company name), which purchases the following chemical(s) from your company:

(List chemicals, products)

OPTION 1: At the time of the visit, (company name) did not have Material Safety Data Sheets (MSDS)/labels for these products despite their prior request for it.

OPTION 2: At the time of the visit, Material Safety Data Sheets (MSDS)/labels supplied by your company were found to be deficient. (Describe the specific deficiencies.)

You are required under OSHA's Hazard Communication Standard (29 CFR 1910.1200) or your State's right-to-know law to perform hazard determinations, label containers, and provide the MSDS for all hazardous chemicals which you produce or import. A copy of the standard is provided for your reference. Please immediately send properly completed material safety data sheets/labels for the chemicals listed above to your customer and a copy to me. If this information is not received within 30 days, an inspection of your establishment may be conducted.

If the MSDS/label described above was deficient, you are also required to send revised copies to all of your customers with the first shipment after a MSDS/label is revised.

Thank you for your assistance. If you have any questions regarding this matter, please feel free to contact me at (insert telephone number).

Sincerely,

Area Director

Appendix C

Hazard Evaluation Procedures

The hazard evaluation procedures required by the standard are performance-oriented. Basically, OSHA's concern is that the information on labels and data sheets, and in the training program, is adequate and accurate. Although specific procedures to follow and number of sources to be consulted cannot be established, general guidance can be provided. The hazard evaluation process can be characterized as a "tiered" approach--the extent to which a chemical must be evaluated depends to a large degree upon the common knowledge regarding the chemical, whether its health effects are under review, and how prevalent it is in the workplace.

1. The first step for CSHO's evaluating chemicals is to determine whether the chemical is part of the "floor" of chemicals to be considered hazardous in all situations.

    a. The floor of chemicals consists of three sources. They are as follows:

    (1) Any substance for which OSHA has a permissible exposure limit (PEL) in 1910.1000, or a comprehensive substance-specific standard in Subpart Z. This includes any compound of such substances where OSHA would sample to determine compliance with the PEL.

    (2) Any substance for which the American Conference of Governmental Industrial Hygienists (ACGIH) has a Threshold Limit Value (TLV) in the latest edition of their annual list is to be included in the Hazard Communication Program. Any mixture or combination of these substances would also be included.

    (3) Any substance which the National Toxicology Program (NTP) or the International Agency for Research on Cancer (IARC) has found to be a suspect or confirmed carcinogen or which OSHA regulates as a carcinogen is to be included in the Hazard Communication Program.

b. Sources to generally establish hazards of the chemicals that are part of the floor of hazardous chemicals covered by the standard:

>The OSHA Chemical Information Manual, OSHA Instruction CPL 2-2.43, October 20, 1987.
>
>NIOSH/OSHA Occupational Health Guidelines.
>
>Documentation for the Threshold Limit Values.
>
>NTP Summary of the Annual Report on Carcinogens.
>
>IARC Monographs.

In addition, the CSHO should check the <u>NIOSH Registry of Toxic Effects of Chemical Substances (RTECS)</u> to see if any hazards are indicated which do not appear in these sources. If there are, further study should be done to evaluate the hazards. RTECS should never be considered a definitive source for establishing a hazard since it consists of data that has not been evaluated. It is, however, a useful screening resource.

2. The second step is to consult other generally available sources to see what has been published regarding the chemical. Patty's <u>Industrial Hygiene and Toxicology</u> would be one such source. OCIS contains a number of other chemical information sources. Material Safety Data Sheets available through information services would also be useful.

3. The third step, for those chemicals where information is not readily available or where such available information is not complete, is to perform searches of bibliographic data bases. In general, the National Library of Medicine (NLM) services should be used. These include the Toxicology Data Bank (TDB), TOXLINE, and MEDLARS. The information generated by these data bases should be evaluated using the criteria in Appendix B of the HCS; i.e., to qualify as an acceptable study, it must be conducted according to scientific principles (e.g., in animal studies, number of subjects is adequate to do statistical analyses of the results; control group is used, and the study must show statistically significant

results indicating an adverse health effect). This evaluation obviously requires a subjective, professional assessment. Any questions should be referred to the Directorate of Compliance Programs, Office of Health Compliance Assistance (through the Regional Office) for assistance. In general, uncorroborated case reports and in vitro studies, such as Ames tests, are useful pieces of information, but not definitive findings of hazards. Animal studies involving species other than those indicated in the acute hazard definitions must be evaluated as well. The acute hazard definitions are not included in the standard to "categorize" chemicals but rather to establish that chemicals meeting those definitions fall under the coverage of the standard.

4. In some cases, the only information available on a substance may be employer-generated data. If the employer indicates that such information is the basis for the hazard evaluation, the CSHO shall ask to see it to complete the OSHA evaluation.

5. In cases where the employer denies the CSHO access to its own hazard data and no published data on the chemical can be found to review the sufficiency of the hazard determination, the Regional Office shall be contacted for assistance in obtaining an administrative subpoena. The Directorate of Compliance Programs shall be contacted if assistance is required in order to obtain unpublished chemical hazard information available from other Federal agencies such as Environmental Protection Agency.

6. If an employer has found any chemical to be non-hazardous, and the CSHO has reason to believe it is hazardous, further investigation is required. The definitions of hazard in the standard are very broad, and it is not expected that many chemicals can be considered nonhazardous under this approach. Those most likely to be exempted would be chemicals that pose no physical hazards, and which have lethal dose findings above the limits found in the acute hazard definitions.

7. In some cases, the employer may not have addressed in the Hazard Communication Program a specific chemical that the CSHO knows to be present through knowledge of the process or through sampling or other investigation of the workplace. This situation should also be further investigated. If the CSHO has information to indicate

that there is a hazard, the employer must be able to defend the finding of no hazard.

Appendix D

Guide for Reviewing MSDS Completeness

NOTE: This guide has been developed for use as an optional aid during inspections.

During CSHO review for Material Safety Data Sheet completeness, the following questions may be helpful:

1. Do chemical manufacturers and importers have an MSDS for each hazardous chemical produced or imported into the United States?

2. Do employers have an MSDS for each hazardous chemical used?

3. Is each MSDS in at least English?

4. Does each MSDS contain at least the:

    (a) Identity used on the label?

    (b) Chemical and common name(s) for single substance hazardous chemicals?

    (c) For mixtures tested as a whole:

    　　(1) Chemical and common name(s) of the ingredients which contribute to the known hazards?

    　　(2) Common name(s) of the mixture itself?

    (d) For mixtures not tested as a whole:

    　　(1) Chemical and common name(s) of all ingredients which are health hazards (1 percent concentration or greater), including carcinogens (0.1 percent concentration or greater)?

    　　(2) Chemical and common name(s) of all ingredients which are health hazards and present a risk to employees, even though they are present in the

mixture in concentrations of less than 1 percent or 0.1 percent for carcinogens?

(e) Chemical and common name(s) of all ingredients which have been determined to present a physical hazard when present in the mixture?

(f) Physical and chemical characteristics of the hazardous chemical (vapor pressure, flash point, etc.)?

(g) Physical hazards of the hazardous chemical including the potential for fire, explosion, and reactivity?

(h) Health hazards of the hazardous chemical (including signs and symptoms and medical conditions aggravated)?

(i) Primary routes of entry?

(j) OSHA permissible exposure limit (PEL)? The American Conference of Governmental Industrial Hygienists (ACGIH) Threshold Limit Value (TLV)? Other exposure limit(s) (including ceiling and other short term limits)?

(k) Information on carcinogen listings (reference OSHA regulated carcinogens, those indicated in the National Toxicology Program (NTP) Annual Report on Carcinogens and/or those listed by the International Agency for Research on Carcinogens (IARC))?

NOTE: Negative conclusions regarding carcinogenicity, or the fact that there is no information, do not have to be reported unless there is a specific space or blank for carcinogenicity on the form.

(l) Generally applicable procedures and precautions for safe handling and use of the chemical (hygienic practices, maintenance and spill procedures)?

(m) Generally applicable control measures (engineering controls, work practices and personal protective equipment)?

(n) Pertinent emergency and first aid procedures?

(o) Date that the MSDS was prepared or the date of the last change?

(p) Name, address and telephone number of the responsible party?

5. Are all sections of the MSDS completed?

INDEX

Appropriate Hazard Warnings 25, 137
Article 20, 119, 120, 121
Carcinogen Labeling 138, 141
Checklist for Compliance 42
Chemical Information List 8
Chemical Manufacturers 20, 23, 25, 121
Chemicals in Unlabeled Pipes 14
Civil Liability 78
Compliance Safety and Health Officers (CSHOs)
 .... 88, 91, 93, 94, 98-103, 105-108,
 133, 137, 147, 153, 157
Container 20, 122
Contractor Employees 5
Copy of Hazard Communication Program ... 18
Employee Information and Training
 .... 13, 28, 41, 60, 62, 5, 105, 148
Evaluation 113
Federal Right-To-Know Law 75
Hazard Communication Coverage 48
 Employee Information and Training
 ..... 13, 28, 41, 60, 62, 95, 105, 148
 General Company Policy 2
 Guidelines for Compliance 36
 Hazardous Chemicals 22, 24, 26, 38
 Hazard Determination 23, 33, 71, 90, 123
 Hazard Evaluation Procedures 153
 Health Hazard Definitions 31
 Hazard Warning 22
 Instructions vi
 Labels 3, 11, 19, 22, 24, 25, 39, 63-65,
 67, 94, 98, 132
 List of Hazardous Chemicals 2, 8
 Material Safety Data Sheets (MSDS)
 .... 3, 9, 10, 22, 40, 51, 53-58, 99,
 101, 142, 151

Non-routine tasks 4, 14
Origin and Purpose 49
Questionnaire 7
Scope and Application 18, 89, 115
Written Hazard Communication Program
 .... 24, 39, 68, 93
Incoming Containers 11
In-plant Containers 11, 137
Labeling 3, 11, 19, 22, 24, 25, 39, 63-65,
 67, 94, 98, 132
Laboratories 18
 Exposure to Hazardous Chemicals .. 113, 117
Manufacturers 12
Medical Records Access 111
Material Safety Data Sheets (MSDS)
 ... 3, 9, 10, 22, 25, 28, 40, 51, 53-58,
 94, 99, 101, 142, 151, 157
Multi-Employer Workplaces
 ...... 24, 97, 104, 128, 129, 131, 147
NIOSH 112, 116, 132, 154
Outside Contractors 69
Preemption of Right-To-Know Laws 73
Safety and Health Manager 2, 4, 5
Secretary of Labor vs. United Steelworkers of
 America 87
Target Organ Effects 32
Trade Secrets 22, 28, 35, 72, 108, 150
Training Hazard Communication
 ...... 13, 28, 41, 60, 62, 95, 105, 148
Voluntary Training Guidelines 95, 106
Written Hazard Communication Program
 .... 24, 39, 68, 93, 127
Written Hazard Determination Program .. 15, 71

About Government Institutes

Government Institutes, Inc. was founded in 1973 to provide continuing education and practical information for your professional development. Specializing in environmental, health and safety concerns, we recognize that you face unique challenges presented by the ever-increasing number of new laws and regulations and the rapid evolution of new technologies, methods and markets.

Our information and continuing education efforts include a Videotape Distribution Service, over 200 courses held nation-wide throughout the year, and over 250 publications, making us the world's largest publisher in these areas.

Government Institutes, Inc.
4 Research Place, Suite 200
Rockville, MD 20850
(301) 921-2355

Other related books published by Government Institutes:

Health Effects of Toxic Substances   This comprehensive book provides you with an excellent understanding of the toxicology and industrial hygiene of hazardous materials. Chapters cover: Industrial Toxicology - History and Hazards; Exposure and Entry Routes - Pharmacokinetics I; Distribution, Localization, Biotransformation, Elimination; Dose-Effects and Time-Effects Relationships; Classification, Type, and Limits of Exposure; Action of Toxic Substances Pharmacodynamics; Target Organ Effects; Reproductive Toxins, Mutagens, and Carcinogens; Survey of Common Hazardous Agents I, Toxic Substances; Survey of Common Hazardous Agents II, Physical & Biological Hazards; Types of Environmental Health Hazards; Monitoring of Harmful Agents; Exposure Limits and Personal Protective Equipment; Exposure Control Methods; Medical Monitoring, Treatment, and Management; Risk Assessment. *Softcover, Index, 300 pages, Aug. '95, ISBN: 0-86587-471-9 $39*

Understanding Workers' Compensation: *A Guide for Safety and Health Professionals*   This book explains in simple and direct terms the Workers' Compensation System. It provides a basic understanding of injury prevention, types of injuries, and cost containment strategies. This book includes sample forms, checklists for work site evaluations, and an appendix containing material from the most recent U.S. Chamber of Commerce analysis, comparisons of all state and Canadian provincial laws, policies on rehabilitation, statistics on benefits payable, and waiting periods. A directory of state and provincial workers compensation administrators with full contact information is also included. *Softcover, 192 pages, June '95, ISBN: 0-86587-464-6 $45*

So You're the Safety Director: An Introduction to Loss Control and Safety Management   Author Michael Manning, a safety veteran and sought-after consultant and speaker, has created an introduction to your bottom-line responsibilities, concentrating on your role in evaluating, managing, and controlling your company's losses and handling the OSHA compliance process. Manning's narrative approach and easy-to-follow writing style make it seem like you've hired him to help you start — or upgrade — your safety program, which is exactly what hundreds of companies have done. Let Manning walk you through the in's and out's of establishing and evaluating your company's safety program: comparing your safety program to those of similar companies, establishing safety committees, involving all employees in your safety program, investigating accidents and preventing their recurrence, managing your compensation costs, preparing for and handling OSHA inspections, and using your company's insurance company as a resource. *Softcover/Index/184 pages/Oct '95/$45   ISBN: 0-86587-481-6*

Ergonomic Problems in the Workplace:A Guide to Effective Management   The valuable insights you'll gain from this new book will help you develop and implement your own successful ergonomics program. Now your company can reduce injuries — such as Cumulative Trauma Disorders (CTD) — and reduce the number of workers' compensation claims. In addition, case studies help you learn from the successes and failures of other companies. Table of contents includes: developing an ergonomics program; management commitment; case histories; hazard assessment; cumulative trauma disorders; workplace hazards; hazard prevention and controls; back injuries and material handling; tool selection; ergonomic personal protective equipment; implementing an ergonomics program; medical management; VDTs and office ergonomics; heat stress; training; ADA and ergonomics; working with OSHA on ergonomic issues; and sources of information and assistance. *Softcover/272 pages/Sept '95/$59 ISBN: 0-86587-474-3*

Call the above number for our current book/video catalog and course schedule.

Publications (cont.)

OSHA Field Inspection Reference Manual — This new revision of inspection guidelines, previously contained in the OSHA Field Operations Manual, is now being used by OSHA inspectors when checking your facility for compliance. Learn where the inspectors will look, what they'll look for, how they'll evaluate your working conditions, and how they'll actually proceed once inside your facility. *Softcover, 144 pages, Jan '95, ISBN: 0-86587-426-3* **$59**

OSHA Technical Manual, 3rd Edition — This OSHA inspection manual includes chapters on: Personal Sampling Techniques and Procedures for Air Contaminants; Sampling for Surface Contamination; Heat Stress; Noise Measurement; Back Disorders and Injuries in Industry; Indoor Air Quality Investigations; Hospital Investigations: Health Hazards; Technical Equipment for Testing and Monitoring; Shipping and Handling of Samples; Pressure Vessel Guidelines; Demolition; Chemical Protective Clothing; Oilwell Derrick Stability; Industrial Robots and Robot System Safety; and more. *Softcover, 300 pages, Nov '93, ISBN: 0-86587-366-6* **$79**

Safety Made Easy: A Checklist Approach to OSHA Compliance Written by Tex Davis, this book provides a new, simpler way of understanding your requirements under the complex maze of OSHA's workplace safety and health regulations. The easy-to-use format and logical organization make this book ideal for those who are just entering the field of safety compliance as well as for experienced safety professionals. *Softcover, 180 pages, May '95, ISBN: 0-86587-463-8* **$45**

Written Compliance Programs are the cornerstone of compliance with OSHA standards, and are always requested by OSHA inspectors. Creating them from scratch is a laborious task, but by using the Wordperfect diskette in this book-disk package, users can customize these boilerplate programs and produce their own company-specific written program, quickly and easily.

Electrical Safety and Lockout/Tagout: *Proven Written Programs for Compliance*
Details what types of work are covered and excluded, compliance procedures, and training.   *Softcover w/ WordPerfect disk, approx. 170 pages, Nov '95, ISBN: 0-86587-502-2* **$59**

OSHA's Respiratory Protection Standard: A Proven Written Program for Compliance
Explains the different types of respirators, maintenance requirements, respirator selection criteria, fit test instructions, and training requirements. It also includes a medical questionnaire and respiratory training record. *Softcover w/ WordPerfect disk, approx. 150 pages, Nov '95, ISBN: 0-86587-501-4* **$59**

OSHA's Process Safety Management Standard: *A Proven Written Program for Compliance*
Features 12 easy-to-use checklists for performing a self-audit of your compliance status, and a convenient, reader-friendly list of the Toxic and Reactive Chemicals regulated by the standard together with the amount of each that must be on hand in order to trigger the standard's coverage. *Softcover w/ WordPerfect disk, approx. 150 pages, Nov '95, ISBN: 0-86587-500-6* **$59**

Educational Programs

■ Our **COURSES** combine the legal, regulatory, technical, and management aspects of today's key environmental, safety and health issues — such as environmental laws and regulations, environmental management, pollution prevention, OSHA and many other topics. We bring together the leading authorities from industry, business and government to shed light on the problems and challenges you face each day. Please call our Education Department at (301) 921-2345 for more information!

■ Our **TRAINING CONSULTING GROUP** can help audit your ES&H training, develop an ES&H training plan, and customize on-site training courses. Our proven and successful ES&H training courses are customized to fit your organizational and industry needs. Your employees learn key environmental concepts and strategies at a convenient location for 30% of the cost to send them to non-customized, off-site courses. Please call our Training Consulting Group at (301) 921-2366 for more information!

License Agreement

By opening this package you indicate your acceptance of the **Government Institutes, Inc.** software licensing agreement with the following terms:

Single Users: For each licensed use of the Software which you have purchased, only one person may access the Software at any given time.

Restrictions: You may not and you may not permit others to (a) disassemble, decompile or otherwise derive source code from the Software, (b) reverse engineer the Software, (c) modify or prepare derivative works of the Software, (d) copy the Software, except to make a single copy for archival purposes only, (e) rent or lease the Software, (f) use the Software in an online system, (g) use the Software in any manner that infringes the intellectual property or other rights of another party, or (g) transfer the Software or any copy thereof to another party, unless you transfer all media and written materials in this package and retain no copies of the Software (including prior versions of the Software) for your own use.

Limited Warranty and Limitation of Liability: For a period of 60 days from the date the Software is acquired by you, Government Institutes warrants that the media upon which the Software resides will be free of defects that prevent you from loading the Software on your computer. Government Institutes' sole obligation under this warranty is to replace any defective media, provided that you have given Government Institutes notice of the defect within such 60-day period. The Software and data are licensed to you on an **"AS IS"** basis without any warranty of any nature.

Government Institutes disclaims all other warranties, express or implied, including the implied warranties of merchantability and fitness for a particular purpose. Government Institutes shall not be liable for any damage or loss of any kind arising out of or resulting from your possession or use of the software and data (including data loss or corruption). In addition, the authors, editors, and publisher assume no liability of any kind whatsoever resulting from the use of or reliance upon the contents of this product. Regardless of whether such liability is based in tort, contract or otherwise. If the foregoing limitation is held to be unenforceable, Government Institutes' maximum liability to you shall not exceed the amount of the license fees paid by you for the software. The remedies available to you against Government Institutes under this agreement are exclusive. Some states do not allow the limitation or exclusion of implied warranties or liability for incidental or consequential damages, so the above limitations or exclusions may not apply to you.